Yashvir Singh, Nishant Kumar Singh, Abhishek Sharma (Eds.)
Biolubricants

Green Tribology and Tribochemistry

Edited by
Yashvir Singh, Nishant Kumar Singh

Volume 1

Biolubricants

Feedstocks, Catalysts, and Nanotechnology

Edited by
Yashvir Singh, Nishant Kumar Singh, and Abhishek Sharma

DE GRUYTER

Editors
Dr. Yashvir Singh
Associate professor
Department of Mechanical Engineering
Harcourt Butler Technical University,
Kanpur, Uttar Pradesh,
India
yashvirsingh21@gmail.com

Dr. Nishant Kumar Singh
Department of Mechanical Engineering
Harcourt Butler Technical University
Kanpur 208002, Uttar Pradesh
India
nishant.singh78@gmail.com

Prof. Abhishek Sharma
Associate professor
Loknayak Jai Prakash Institute of Technology,
Chapra, Bihar,
India
absk2001@gmail.com

ISBN 978-3-11-114512-9
e-ISBN (PDF) 978-3-11-114513-6
e-ISBN (EPUB) 978-3-11-114555-6
ISSN 2940-0376

Library of Congress Control Number: 2024933405

Bibliographic information published by the Deutsche Nationalbibliothek
The Deutsche Nationalbibliothek lists this publication in the Deutsche Nationalbibliografie; detailed bibliographic data are available on the Internet at http://dnb.dnb.de.

Cover image: kadmy/iStock/Getty Images Plus
Typesetting: Integra Software Services Pvt. Ltd.
Printing and binding: CPI books GmbH, Leck

www.degruyter.com

Preface

Lubricants serve as an antifriction medium and lessen wear on the parts that come into contact with one another when machine parts move. Petroleum-based reserves are running out worldwide, which might lead to price increases and raise concerns about environmental damage. Biolubricants are an alternative option to mineral oils as they are biodegradable and have natural technical qualities. The suitability of feedstocks has constraints in terms of appropriateness because of contemporary issues such as their role in the human food chain and the environmental harm that results from using available cultivation land. Biolubricants, feedstocks, catalysts, and nanotechnology are comprehensive, interdisciplinary, and timely solutions to this subject of concern. Chapter 1 discusses eco-friendly routes for producing biolubricants. Chapter 2 suggests vegetable oils as suitable feedstocks for the production of biolubricants with the introduction of catalysts. Chapter 3 reviews the opening of castor oil-based biolubricants and their applications. Chapter 4 introduces biodegradable biolubricants and submitted their appropriate applications. Chapter 5 shows the suitability of nonedible vegetable feedstocks for developing biolubricants. Chapter 6 describes the alternative fuels and their usage as biolubricant applications. Chapter 7 discusses the usage of activated biochar and its environmental and agricultural applications. Chapter 8 focuses on solar-based phase change material applications related to environmental usage. Chapter 9 reviews the application of biomass, municipal waste, and agriculture waste for the production of biolubricants. Chapter 10 focuses on the proper feedstocks for the generation of biolubricants and their challenges in the future. Chapter 11 presents a life cycle analysis and imparts knowledge related to their degradability. Chapter 12 discusses the problems associated with the production of biolubricants and how they can be resolved by implementing nanotechnology in this field.

https://doi.org/10.1515/9783111145136-202

Contents

List of authors

Chandu S. Madankar
Department of Oils, Oleochemicals
and Surfactants Technology
Institute of Chemical Technology
Matunga, Mumbai
Maharashtra, India
chandumadankar@gmail.com
Chapter 1

Priscilla Agnel
Department of Oils, Oleochemicals
and Surfactants Technology
Institute of Chemical Technology
Matunga, Mumbai
Maharashtra, India
oil22p.agnel@pg.ictmumbai.edu.in
Chapter 1

Dr. Sergio Nogales-Delgado
Department of Applied Physics
University of Extremadura
Avda. De Elvas s/n
06006 Badajoz, Spain
senogalesd@unex.es
Chapter 2

Prof. Dr. Juan Félix González
Department of Applied Physics
University of Extremadura
Avda. De Elvas s/n
06006 Badajoz, Spain
jfelixgg@unex.es
Chapter 2

Prof. Dr. José María Encinar
Department of Chemical Engineering
and Physical Chemistry
Avda. De Elvas s/n
06006 Badajoz, Spain
jencinar@unex.es
Chapter 2

Dr. Lokesh Kumar
Dr. S.S. Bhatnagar University Institute
of Chemical Engineering
Panjab University
Chandigarh 160014, India
mrtlokesh125@gmail.com
Chapters 3, 4

Anupama
Department of Chemistry
Dhanauri P.G. College
Dhanauri, Haridwar 247667
Uttarakhand, India
anupamapawar85@gmail.com
Chapter 3

Dr. Dheeraj Ahuja
Chemical Engineering Department
Motilal Nehru National Institute of Technology
Teliyarganj, Prayagraj 211004
Uttar Pradesh, India
dheerajahuja@mnnit.ac.in
Chapter 3 and 4

Dr. Sheetal Monga
Dr. S.S. Bhatnagar University Institute
of Chemical Engineering
Panjab University
Chandigarh 160014, India
monga.sheetal85@gmail.com
Chapter 4

Dr. Anupama Kaushik
Dr. S.S. Bhatnagar University Institute
of Chemical Engineering
Panjab University
Chandigarh 160014, India
and
Coordinator, Energy Research Centre
Panjab University
Chandigarh, India
anupamachem@gmail.com
Chapter 4

https://doi.org/10.1515/9783111145136-204

J. Prabhakaran
Renewable Energy and Alternative Fuel Laboratory
Department of Mechanical Engineering
National Institute of Technology, Srinagar
Jammu and Kashmir, India
prabhakaran.pusa@gmail.com
Chapter 5

Amit Kumar
Renewable Energy and Alternative Fuel Laboratory
Department of Mechanical Engineering
National Institute of Technology, Srinagar
Jammu and Kashmir, India
amitskjaipur@gmail.com
Chapter 5

Mohd. Mujtaba Ahmed
Renewable Energy and Alternative Fuel Laboratory
Department of Mechanical Engineering
National Institute of Technology, Srinagar
Jammu and Kashmir, India
mohdmujtabaahmed5@gmail.com
Chapter 5

Harveer Singh Pali
Renewable Energy and Alternative Fuel Laboratory
Department of Mechanical Engineering
National Institute of Technology, Srinagar
Jammu and Kashmir, India
hspali@nitsri.ac.in
Chapter 5

Priyanka Singh
Department of Mechanical and Automation Engineering
Indira Gandhi Delhi Technical University for Women (IGDTUW)
New Delhi, India
pshetu0@gmail.com
Chapter 6

Nathi Ram Chauhan
Department of Mechanical and Automation Engineering
Indira Gandhi Delhi Technical University for Women (IGDTUW)
New Delhi, India
nrchauhan@igdtuw.ac.in
Chapter 6

Dr. J. P. Yadav
Department of Mechanical Engineering
Baba Saheb Dr. Bhimrao Ambedkar College of Agricultural Engineering and Technology
Etawah 206001
Uttar Pradesh, India
jp1890@rediffmail.com
Chapter 7

Mr. Pratyush
Department Civil Engineering
Kamla Nehru Institute of Technology
Sultanpur
Uttar Pradesh, India
pratyushjeet067@gmail.com
Chapter 7

Dr. Devendra Singh
Department of Mechanical Engineering
Faculty of Agricultural Engineering and Technology
Chandra Shekhar Azad University of Agriculture and Technology, Kanpur, Campus Etawah
Uttar Pradesh, India
devendra.17@rediffmail.com
Chapter 7 and 8

Dr. Ajay Kumar Sharma
Department of Mechanical Engineering
Institute of Engineering and Technology
Lucknow 226021
Uttar Pradesh, India
ajaysharma@ietlucknow.ac.in
Chapter 8

Shubham Mishra
Department of Mechanical Engineering
Institute of Engineering and Technology
Lucknow 226021
Uttar Pradesh, India
shubhammishra13673@gmail.com
Chapter 8

Rakesh Pant
Department of Microbiology
Graphic Era (Deemed to be) University
Dehradun 248001, Uttarakhand, India
And
Department of Quality
Eureka Forbes Ltd., Lal Tappar
Dehradun 248140, Uttarakhand, India
rakesh.microbiologist@gmail.com
Chapter 9

Khushi Sah
Department of Microbiology
Graphic Era (Deemed to be) University
Dehradun 248001, Uttarakhand, India
sahkhushi2003@gmail.com
Chapter 9

Nirmal Patrick
Department of Quality
Eureka Forbes Ltd., Lal Tappar
Dehradun 248140, Uttarakhand, India
nirmalpatrick@gmail.com
Chapter 9

Amit Gupta
Department of Zoology
University of Jammu
Jammu 180006, Jammu and Kashmir, India
amitrrl@gmail.com
Chapter 9

Anil Dhanola
Department of Mechanical Engineering
University Institute of Engineering
Chandigarh University
Gharuan (Mohali) 140413, Punjab, India
anildhanola3698@gmail.com
Chapter 10

Chandra Mohan
Chemistry, SBAS,
K. R. Mangalam University
Gurugram 122103, Haryana, India
gurgaonmohan@yahoo.co.in
Chapter 11

Jenifer Robinson
Department of Science and French
Indian School Al Wadi Al Kabir
Muscat, Sultanate of Oman, Oman
iswkscience@gmail.com
Chapter 11

Rekha Goswami
Department of Environmental Science
Graphic Era Hill University
Dehradun, Uttarakhand, India
rgoswami.2324@gmail.com
Chapter 12

Barkha Kamal
Department of Botany
DBS PG College
Dehradun, Uttarakhand, India
dayabrij@gmail.com
Chapter 12

Abhilasha Mishra
Department of Chemistry
Graphic Era (Deemed to be University)
Dehradun, Uttarakhand, India
abhi1680@gmail.com
Chapter 12

Chandu S. Madankar, Priscilla Agnel

1 Eco-friendly processing routes for the production of biolubricants

1.1 Introduction

Currently, the lubricant industry is dominated by petroleum- and mineral oil-based lubricants, but these lubricants are toxic and are susceptible to loss, thereby reaching the environment and affecting the living organisms. On the other hand, biolubricants are nontoxic, biodegradable, and sustainable; when compared to petroleum-based lubricants, the loss to the environment is less. Replacing petroleum-based oils with vegetable-based oils is one of the ways to reduce the adverse effects on the environment.

Lubricants can be broadly classified into five groups according to the API system: Groups I–III include lubricants derived from petroleum products and Group III oils are called synthetics. Group IV contains highly processed base oils consisting of saturated polyalphaolefins and has desirable properties, including a wider temperature range. Group V oils do not fit into Groups I–IV and include synthetics like polyalkylene glycols (APG), monoesters, diesters, and polyesters. Biolubricants fall into Group V class. Biolubricants can be produced using triglycerides, fatty acids, and fatty acid methyl esters by means of chemical modification. Transesterification/esterification, oligomerization (to produce esters from dimer and trimer acids), epoxidation, estolide formation (this requires capping fatty acid, which makes the overall production cost expensive), and so on. Some other chemical modifications include metathesis, hydroformylation, acetylation, and so on.

Recent emphasis on sustainability has led to a paradigm shift toward “clean and green” processes. To render, the existing manufacturing processes more effective and less expensive, reworking and switching to new techniques is essential.

There is an increasing need to identify alternative methods for manufacturing because environment-friendly and sustainable processes and technologies, such as using biomass as industrial raw material and using biomolecules and other bioactive agents for catalysis, are feasible alternatives to conventional methods. With the world becoming overly conscious of carbon emissions and footprints, along with efforts to meet increasing global demands, there arises a need to develop eco-friendly processes. An ideal production process would be environmental friendly, would require minimal input, and would be economically feasible. Eco-friendly processing reduces the energy requirement for the overall process and works toward less toxic reaction pathways and minimal effluents as shown in Fig. 1.1.

Plant oils can be employed as lubricants. Plant oils are superior to mineral oils in terms of lubricity; they exhibit outstanding lubricity. Additionally, the viscosity index

https://doi.org/10.1515/9783111145136-001

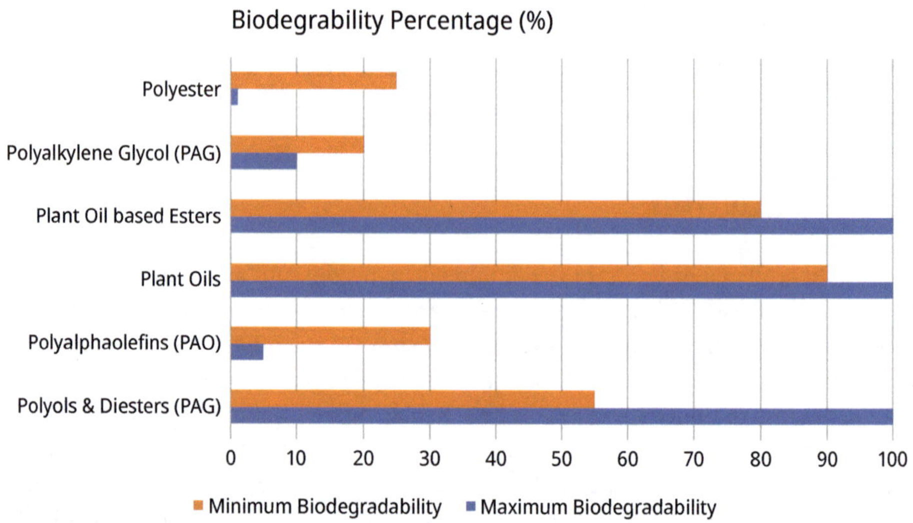

Fig. 1.1: Biodegradability of different classes of lubricants [1].

(VI) of plant oils is quite high. Plant oils have a very high VI and high flash points. They are nontoxic unlike mineral oils, are readily biodegradable, and are obtained from renewable sources.

1.2 Electrochemical decarboxylation via Kolbe electrolysis

The Kolbe electrolysis (KE) is also known as "anodic decarboxylation," which was named after Hermann Kolbe. The Kolbe reaction is essentially known as the decarboxylation of fatty acids to produce dimers. KE is one of the oldest electroorganic syntheses. Michael Faraday was the first one to observe this when he studied the electrolysis of aqueous acetic acid in the year 1834 [2]. By electrolyzing salts of carboxylic acids, Kolbe created salicylic acid; this procedure is sometimes referred to as the "Kolbe–Schmitt reaction." Hydroisomerization and KE are the final steps in the synthesis of lubricants, and the end product is isoalkanes.

1.2.1 Reaction mechanism

Fatty acids or sodium salts of fatty acids undergo electrolysis to form carboxylate ions; on oxidation at the anode, carbon dioxide is evolved, leaving behind alkyl-free radicals to form dimers. This reaction mechanism is supported by Kolbe's "active oxygen theory," which was continued by the "peroxide theory" of C. Shall in the year 1896 [3].

The general scheme of the reaction mechanism is shown in Fig. 1.2. The mechanism shows the formation of main product, that is, Kolbe product dimer and non-Kolbe (alcohols, ethers, alkenes, esters) as side products.

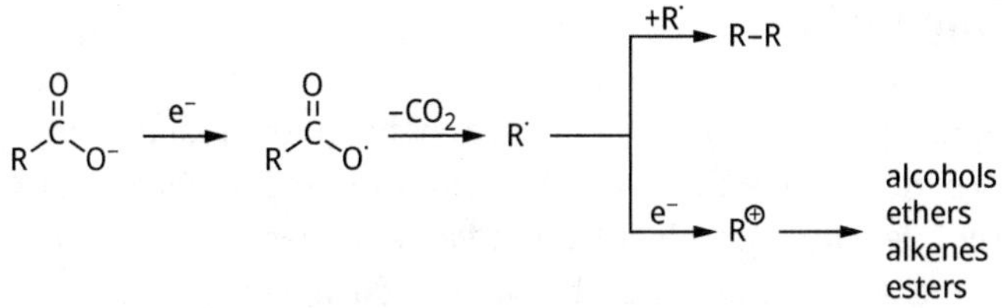

Fig. 1.2: Reaction mechanism of Kolbe electrolysis method [3].

It is stated that acetic acid undergoes anodic oxidation to form ethane (or any alkane in that case is termed as "Kolbe dimer") or methyl alcohol (Hofer–Moest product). Methyl alcohol is a higher homolog obtained by anodic oxidation at a lower potential than that required for the formation of ethane [4].

Raw materials that are used for the process can be petroleum or bio-based sources such as algae, plants, or animals. Feedstock is not limited to triglycerides; partially processed oils containing monoglycerides and diglycerides can also be included in this group. Substitution of petroleum-based mineral oils by biobased oils greatly reduces dependency on petroleum and reduces environmental impact [5]. Other feedstocks that can be used, but are not limited to, are plant oil, plant fat, plant waxes, animal oils/waxes, and fish oils/waxes. Plant oils that can be used include edible vegetable oils such as canola, cottonseed, linseed, palm kernel, palm stearin, rapeseed, rice bran, sesame, sunflower, and so on. Nonedible oils, such as jatropha, castor, karanja, and high oleic oils obtained from bioengineered plants can also be thought of for the purpose. Examples of animal fat include tallow, lard, and fish oil. Certain strains of algae, especially microalgae, may contain high levels of lipids proportionally. Algal sources for the algae oils can contain from 2 wt% to 40 wt% of lipids, based on the total weight of the biomass [6]. Fatty acids obtained by hydrolysis, transesterification, or pyrolysis, fatty esters obtained by transesterification, and fatty alcohols obtained as reduction products and waste and recycled food-grade fats and oils are also other reliable feedstocks for the process [6] as shown in Fig. 1.3.

At Anode:	$2\,RCOO^-$	+	H_2O	→	$2\,CH_3$	+	$2\,CO_2$			
At Cathode:			$2\,H_2O$	→	$2\,H^+$	+	$2\,OH^-$			
Overall Reaction:	$2\,RCOOH$	+	$2\,H_2O$	→	$2\,C_2H_6$	+	H_2		+	$2\,H_2O$

Fig. 1.3: Reactions at the electrodes in the Kolbe electrolysis process.

1.2.2 Factors influencing yield

The yield and selectivity of Kolbe products, that is, dimers depend on the reaction conditions and the chemical structure of the fatty acid [3]. The influencing factors are temperature, pressure, current density, pH, solvent, additives, and electrode material. As electrolysis proceeds, there is a formation of carboxylate radical, which is proportional to current density and carboxylate concentration. High radical concentration at the electrode surface promotes dimerization. Above the critical potential, Kolbe dimerization proceeds smoothly, which is about 2.4 V (vs NHE) [7]. The oxygen evolution and solvent oxidation are effectively suppressed at this potential in aqueous and nonaqueous solvents [8]. It is assumed that at the critical potential, the carboxylate forms a layer at the electrode surface, wherein the solvent is desorbed. And oxidation of solvent takes place at conditions below the critical potential [3]. For the Kolbe reaction, a neutral pH or a moderately acidic medium is preferred. This is accomplished by utilizing alkoxides to neutralize the carboxylate to an extent of 2–5%. As the reaction proceeds, carboxylate ions are consumed in the anode, and the base is liberated continuously from the cathode. Therefore, the transition of the electrolyte to an alkaline pH can effectively serve as an indicator of the process' completion. This process is known as the "salt-deficit method." In cases where the complete conversion of carboxylic acids to carboxylate ions is essential to improve solubility in an aqueous medium, a mercury cathode is used in the Dinh–Nyugen cell [9]. Due to the formation of amalgams on discharge of alkali metals, the electrolyte is maintained at a neutral pH.

High pressure in aqueous solutions enhances Kolbe coupling over non-Kolbe products [10]. When methanol is utilized as a solvent, transesterification is the predominant reaction, given the reaction temperature is high. Higher temperatures encourage the conversion of fatty acids to fatty acid methyl esters, which compete with electrolytic decarboxylation. Fatty acid methyl esters may, therefore, take over as the primary product when temperatures are raised above 65 °C [11]. The formation of a carboxylate layer over the electrode prevents methanol from oxidizing, and it is considered the most appropriate solvent for KE [12, 13]. Therefore, for desired product formation, the reaction temperature must be maintained below 65 °C. Polar solvents are ideal for the process of electrolysis, but the Kolbe products are soluble in nonpolar solvents. Therefore, a combination of polar and nonpolar solvents can be utilized for optimum performance in KE [6].

Ethanol, methanol, hexane, toluene, nonane, xylene, decane, acetonitrile, undecane, dodecane, dimethylformamide, and 4-methyl morpholine are a few solvents that can be used in combinations. The addition of electrolytes containing perchlorate salts, such as p-toluene sulfonate salts, could raise the electrical conductivity [6]. A portion of the fatty acids can also be converted into carboxylate salts to serve the purpose of electrolytes, by the addition of alkaline hydroxides (such as sodium hydroxide or potassium hydroxide) before or during the KE reaction [6, 14]. The resultant product phase separates and is collected as a layer either at the top or at the bottom of the reactor [15].

The dimer produced by the KE process is subjected to hydroisomerization procedure to change its characteristics and to make it more appropriate for use as lubricant base oil. Figure 1.4 represents a sequential flow of steps involving the production of isoalkanes from starting triglycerides or fatty acids as raw materials. The catalyst involved is a silica–alumina zeolite with impregnated platinum, and the conditions are as follows: hydrocarbon ratio of 2–50, temperature range of 250–400 °C, and pressure range of 10–400 bar [16].

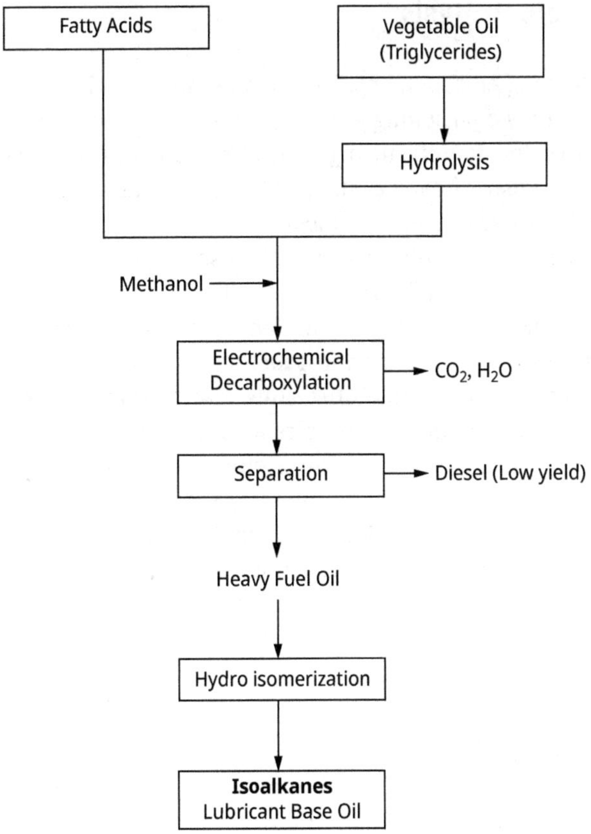

Fig. 1.4: Schematic flowsheet of synthesis of renewable hydrocarbons.

1.2.3 Reaction conditions

Stainless steel, nickel, or graphite can be employed as cathode materials for KE of fatty acids, while titanium with a platinum coating is used as the anode [6]. Carbon materials like graphite, baked carbon, platinum group metals (such as platinum and palladium), or a combination of both have all been mentioned as acceptable electrode materials [6]. A direct current is provided while the electrodes are submerged in the reaction mixture. From 0.05 to 1.0 A/cm^2, current density can be applied [5, 14]. The electrodes' distance from one another can be anything between 1 and 3 mm [6].

Atmospheric pressures and temperatures between 15 and 100 °C, or more frequently, 40–60 °C, are used to perform the KE reaction. A water jacket is used to keep the reactor at the desired temperature [6] depending on the relative densities of the reagent and product.

1.2.4 Advantages of Kolbe electrolysis

KE has a higher potential for achieving environmental benefits because it utilizes renewable and off-peak electricity. Catalyst poisoning is a very common problem when poor-quality raw materials are used for manufacturing. This affects the efficiency of the catalyst and also reduces the number of batches the catalyst can be used, but KE does not experience this issue to the same degree. The moderate operating conditions employed in KE reduce energy consumption while ensuring plant safety.

The thermochemical approach uses more hydrogen than the KE route since it calls for prehydrogenation, ketonization, hydrodeoxygenation (HDO), and hydroisomerization. The processing stages of the thermochemical method rely primarily on catalysts and high temperatures. For hydrogenation of starting materials and hydroisomerization of Kolbe product, the KE method does require a certain amount of hydrogen. By integrating the prehydrogenation and posthydroisomerization procedures into a single pretreatment phase, it may be possible to intensify the KE method and lower the capital costs and hydrogen consumption. KE is a carbon-neutral process [2], and hydrogen and carbon dioxide gases evolved in large amounts during electrolysis are mixed to power to liquid processes for their complete consumption as alternatives to conventional fuels [17, 18].

1.3 Chemoenzymatic synthesis

Industries have recently been concerned about the need for clean, efficient, and high-yielding procedures [19]. Primarily, because of mild reaction conditions, high selectivity and catalytic activity for both natural and synthetic substrates, and high degree of product purity, the use of lipases for product manufacturing is very appealing, proving itself to be an environmental-friendly process [20, 22]. Enzymes often end with the suffix -in (trypsin, chymotrypsin, etc.), and they are also frequently called by their common names that end with -ase and allude to the reaction they catalyze. Oxidoreductases, transferases, hydrolases, lyases, isomerases, and ligases are the six major subcategories in enzyme classification [23]. Enzymes are identified using an Enzyme Commission (EC) number consisting of four digits, separated by a full stop. The EC number came into practice due to the complexity and inconsistency in the nomenclature of enzymes and the use of common names. For instance, lactate dehydrogenase is the common name of the enzyme. According to the nomenclature of enzymes, it is known as L-lactate: NAD^+ oxidoreductase and has the EC number 1.1.1.27 [23].

Lipases (EC 3.1.1.3), also called triacylglycerol ester hydrolyses, catalyze the hydrolysis of triglycerides to glycerol and free fatty acids (FFA) over an oil–water interface [24]. Lipases promote reverse reactions of esterification, interesterification, and transesterification to produce glycerides from glycerol and fatty acids in nonaqueous settings. Lipases, in addition to hydrolysis, also catalyze the acidolysis, alcoholysis, and aminolysis of triglycerides. The three main sources of lipases are all found in nature; they are plants, animals, and microbial sources. Lipases were initially extracted from the pancreas of animals and were utilized as supplements that help in digestion in humans. Microbial lipases from proteases and carboxylases are way more expensive than those obtained from other sources. This is due to the poor yield of the fermentation process [22]. However, further developments in molecular biology have enabled enzyme manufacturers to provide cheaper and high-activity microbial lipases to the market [25]. Currently, microbial lipases are produced by several companies worldwide, including Novozymes, Amano, and Gist Brocades [26].

Enzymes in their free and purest form are not preferred. They have disadvantages, such as high costs and difficulties regarding regeneration. Enzymes adsorb on polar molecules and tend to form aggregates that drastically reduce the catalytic activity of lipases due to nonuniform dispersion in the substrate. Therefore, enzymes in their immobilized form are highly attractive in terms of usage and applications. Lipases can be immobilized by different methods. Some of the immobilization techniques like physical adsorption, covalent bonding, entrapment, cross-linked enzyme aggregate (CLEAs), and encapsulation are discussed in detail by Thangaraj and Soolomon [19]. A suitable immobilization technique is selected based on the following characteristic properties: activity of enzyme, stability, cost, reagent and product toxicological profile, and other favorable characteristics [27].

1.3.1 Enzyme-catalyzed transesterification

A wide range of ester-based oils could potentially be produced by combining readily accessible fatty acids and alcohols. Esters for lubricant applications are chemically categorized as polyesters, oligomeric esters, triglycerides, diesters, and monoesters depending on the number of ester bonds present.

Musa et al. [28] reported the synthesis of trimethylolpropane ester using castor oil methyl ester (COME) as raw material via conventional transesterification reaction. Trimethylol propane (TMP) and KOH catalysts were added to COME after it was heated to 60 °C. The methyl ester to TMP molar ratio was 4:1. The reaction was carried out at 130 °C for 1 h, and the concentration of KOH was 0.8% by weight of COME. Following the completion of the reaction, the organic layer is separated using a separating funnel and rinsed with warm water containing ortho-phosphoric acid, which is equivalent to 10% by weight.

Cavalcanti et al. [29] carried out an enzymatic transesterification reaction between methyl ricinoleate and trimethylolpropane; the reaction was conducted at 40–50 °C with continuous stirring. Reactions were performed at atmospheric pressure; vacuum was applied to facilitate methanol removal to increase the yield. Initially, TMP and methyl ricinoleate were preheated, and lipase was added (1–4% w/w of methyl ricinoleate + TMP mass). The reaction was conducted for 120 h, with samples withdrawn every 24 h. The mixture is centrifuged for 20 min and stored below 0 °C.

Lipases also offer an enzymatic route for the synthesis of estolides [30], epoxides, and dimers, which find their use in a wide range of lubricant applications.

1.3.2 Enzyme-catalyzed dimerization

Dimerization and oligomerization are technologically feasible modifications of the double bonds of unsaturated fatty acids. The process involves transformation of C18 fatty acids with one or more double bonds and continues at 210–250 °C in the presence of acid-activated clay (montmorillonite), resulting in a complex combination. The product is a mixture of monomer, dimer, and trimer fatty acids. The activation process requires the use of strong acids, such as sulfuric acid, which can be costly and potentially hazardous to handle. The amount of acid needed for activation is significant, leading to increased expenses and potential environmental concerns if not handled properly.

Cytochrome P450 enzymes (P450s) belong to the class of oxygenase. Cytochrome P450s can catalyze dimerization reactions based on their catalytic mechanism. The activation of the C–H bond present in fatty acids leads to the formation of carbon-centered radicals [31], whereas in conventional method using acid-activated clay catalyst proceeds through formation of carbenium ions. Laane et al. [32] studied biodimerization of unsaturated fatty acids by using lipoxygenase. Soya-bean lipoxygenase is capable of catalyzing anaerobic formation of dimeric fatty acids from linoleic acid. The maximum substrate concentration with respect to linoleic acid was 9.2 g/L (62% conversion). Theoretically, the maximum overall yield for soya fatty acids is 58%. Monounsaturated and saturated acids are not substrates for lipoxygenase. However, the yield achieved was as low as 3–9% of industrially accepted products.

1.3.3 Advantages and disadvantages of enzymes

The use of enzymes comes with several advantages and disadvantages.

Advantages of enzymes:
- Specificity: Enzymes are highly specific in their substrates and do not interfere with other molecules to produce unwanted compounds.
- Reaction conditions: Enzymes work under mild reaction conditions, typically at lower temperatures and ambient pressures. This not only reduces energy requirements but also minimizes the risk of side reactions or the formation of unwanted by-products.
- Simplicity: Since enzymes have mild reaction conditions, it makes the overall process easier to handle and less hazardous and completes the process in a few steps.
- Biodegradability: Enzymes are generally biodegradable and do not persist in the environment. It reduces the effect on the environment as enzymes break down naturally without causing long-term pollution.
- Product: The final product produced has high purity, due to lack of side reactions or by-product formation. The yield is also high.
- Selectivity in stereochemistry: Enzymes can catalyze reactions with high stereoselectivity; they can produce specific stereoisomers of a molecule.
- Tolerance to FFA: Enzymes have high tolerance for FFA and water. It gives good results even with the use of poor-quality feedstock.
- Compatibility with aqueous systems: Enzymes are generally compatible with aqueous systems, which is advantageous in many industrial processes where water-based solutions are used. This allows for easier handling, separation, and purification of reaction products.
- Separation: Enzymatic reactions are heterogeneous processes, making catalyst separation much easier as compared to homogeneous processes.

Disadvantages of enzymes:
- Cost and production challenges: Enzymes can be expensive to produce and purify, especially when obtained from specialized organisms or through genetic engineering. The production process may require specific infrastructure, equipment, and expertise, thereby increasing the overall production costs.
- Sensitivity to environmental conditions: Enzymes can be highly sensitive to environmental conditions such as temperature, pH, and presence of certain chemicals. They may denature or lose activity outside their optimal conditions, limiting their stability and functionality in certain industrial processes.

- Difficult recovery and reusability: Enzyme recovery and reuse can be challenging due to their complex structure and sensitivity to purification processes. Recycling and reusing enzymes may require additional steps, thus increasing operational complexity and costs.
- Reaction time: Enzyme reactions have very low reaction speed, therefore long reaction times.

1.4 Eco-friendly feedstocks

The search for new feedstock for biolubricants is driven by the desire to achieve environmental sustainability, resource conservation, improved performance, and compliance with market demands and regulatory requirements. The growing demand for sustainable and eco-friendly products, coupled with tightening environmental regulations, acts as the fuel. While biolubricants have made significant strides in terms of performance, there is always a quest for new feedstocks that can further enhance their lubrication properties, stability, and compatibility with different applications. The search for alternative feedstocks allows for the development of biolubricants that can meet or exceed the performance requirements of specific industries and applications.

1.4.1 Waste cooking oil (WCO)

Waste-to-feedstock processes can help mitigate the environmental impacts associated with waste disposal. By diverting WCO to industrial feedstock pollution is minimized and there is a decrease in the overall carbon footprint.

The US produces about 10 million tons annually of used cooking oil, contributing about ~ 55%, followed by China and India contributing about 4.5 and 3.2 million tons annually to the global market [33]. The country with the lowest WCO output is Ireland, which contributes 1% of annual global production of 0.153 million tons per year [34]. Only 3% of WCO is generated annually in Malaysia, according to the data. Palm oil is chosen as the principal source for the manufacture of edible oil in Malaysia as shown in Fig. 1.5.

This was due to the low processing costs and the availability of palm oil generated. The global market of WCO is expected to grow extensively. The global WCO market size that was $5.50 billion in 2019 is estimated to reach $8.48 billion by 2027 [36]; therefore, it is a dependable source of feedstock.

WCOs generally have very high amounts of FFA and, therefore, high acid values. The WCO mainly consists of variable percentage of FFA : oleic (C18:1), linoleic (C18: 2), and linolenic (C18: 3) fatty acids in form of tri-, di-, and monoglycerides [37]. The primary goals of physical treatment of WCO are to remove undesirable components

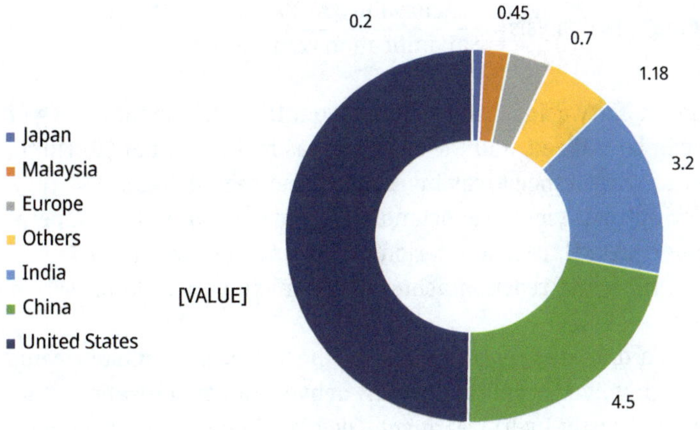

Fig. 1.5: Global WCO production in million tons/year [33–35].

from crude WCOs and to produce oil that is fatty acid enriched. Three physical approaches to WCO regeneration are conceivable to take into consideration, namely, separation using solubility, filtration, and boiling point [38].

$$\underset{\text{Triglyceride}}{C_3H_5(OCOR)_3} + H_2O \xrightarrow[48\text{ bar}]{250\ °C} \underset{\text{Glycerol}}{C_3H_5(OH)_3} + \underset{\text{Fatty Acid}}{3\ RCOOH}$$

$$\underset{\text{Fatty Acid}}{RCOOH} + \underset{\text{Alcohol}}{R'OH} \rightleftharpoons \underset{\text{Ester}}{RCOOR'} + H_2O$$

Fig. 1.6: General reaction involved in hydrolysis of fats/oils and esterification of fatty acids.

Chowdhury [39] studied enzymatic hydrolysis using *Candida rugosa* for the production of octyl esters as the main product using WCO as the raw material. Synthesis involved two steps – enzymatic hydrolysis followed by esterification. The above-mentioned chemical equations represent a general reaction of hydrolysis of triglycerides esterification of fatty acids. Figure 1.6 shows the general reaction involved. Filtered WCO is used as used cooking oil contains impurities and residue. Lipase solution with an enzyme concentration of 1.0 g L^{-1} was added to WCO at 30 °C, and 250 rpm shaker was used to incubate the samples for varied amounts of time ranging from 1 to 30 h for the reaction to attain equilibrium. The reaction proceeds in the presence of water; the range of the water–to-waste-oil ratio is 0.5:1 to 5:1 by volume. FFA are separated using *n*-hexane as the solvent by centrifuging and further distilling to remove the solvent(s). Trace amount of water is removed by the addition of anhydrous sodium sulfate. Percentage hydrolysis is determined by acid value divided by saponification value multiplied by 100:

$$\text{Percentage hydrolysis} = \frac{\text{Acid value (AV)} \times 100}{\text{Saponification value (SV)}} \tag{1.1}$$

The molar ratio of octanol to fatty acids can be altered from 1:1 to 4:1. About 0.2–3 g of Amberlyst 15H catalyst can be utilized, and the reaction was maintained at 60–110 °C. Additionally, the impact of stirrer speed was investigated between 450 and 650 rpm. The ideal parameters for optimal yield are: octanol:FFA molar ratio of 3:1, temperature of 80 °C, 2 g of Amberlyst 15H, reaction period of 3 h, and 50% weight of FFA of desiccant. In the shortest time, the reaction achieved the maximum yield of 98% of octyl esters.

Jahromi [40] proposes a four-step reaction path for the synthesis of biolubricants from WCO. The four steps involved were: hydrolysis, dehydration/ketonization, Friedel Craft (FC) alkylation, and lastly hydrotreatment. Oleic acid and stearic acid were studied as model compounds and WCO was used as actual feedstock. Hydrolysis of WCO was performed under 400 psi N_2 at 250 °C with an oil-to-water mass ratio of 3:1. Dehydration/ketonization process with magnetite as a catalyst was then carried out with a 35:1 feed-to-catalyst ratio and 350 psi cold N_2 pressure. The liquid products were collected to separate oil, catalyst, and leftover solids using a centrifuge for 10 min. The combination of the dehydration and ketonization products was then poured into the Parr reactor. To reduce self-condensation and obtain a 15 wt% total liquid feed, a chosen equimolar combination of COHC was supplied to the reactor as the limiting reactant(s). Cu/ZSM_5-MgO catalyst with a feed-catalyst mass ratio of 30:1 was added to this combination, which was subsequently nitrogen-pressurized to 350 psi.

FC acylation/alkylation process was conducted for 3 h and slowly heated to 80 °C. After cooling to an ambient temperature, the FC product is centrifuged. Then, mild hydrotreatment is done at 200 psi with 10% H_2 (balancing nitrogen) and a Ni/$SiO_2Al_2O_3$ catalyst. After cooling, the liquid by-products were collected using centrifugation. Biolubricant obtained is neutralized with 0.1 M KOH to eliminate unreacted fatty acids that add to the total acid number (TAN), and distilled to remove water and volatile compounds.

Furthermore, Zhang et al. [41] suggested chemically altering WCO with enzymes and ionic liquid to produce branched lubricant in a green manner. In the beginning, WCO was broken down by lipase (*Candida* sp. 99–125) to produce unsaturated fatty acids, which would then be condensed by the urea complex. Then, 2-ethyl hexanol coupled with Novozym 435 was used to convert unsaturated fatty acids to esters. Finally, when the esters were epoxidized with octanoic acid by employing ionic liquid and [HMIm][PF6], octylated branched biolubricant was produced.

1.4.2 Plant biomass

By repurposing waste nonoil-bearing parts of plants, natural resources, such as minerals, fossil fuels, and water, can be conserved, which are often required for the production of raw materials.

The most wide-ranging and renewable resource is lignocellulosic biomass, which is created by photosynthesis [42]. Cellulose, hemicellulose, and lignin are the main constituents of lignocellulosic biomass. Fuels, lubricants, and chemicals can be manufactured from alcohols, aldehydes, carboxylic acids, furfurals, and ketones, obtained from C5 and C6 sugars, acetone–butanol–ethanol (ABE) fermentation combinations, and derivatives of fatty acids and other carboxylic acids obtained from biomass [43].

Lui et al. [44] reported a 70% yield of C_5–C_6 sugars from *Pennisetum* raw biomass in 2.5 h via hydrolytic hydrogenation using solid acid and Ru/C catalyst mixture. Biomass was oven-dried at 105 °C to remove the moisture and was passed through 40–60 mesh. Biomass, solid acid, that is, ZrP and 5 wt% Ru/C catalyst, is used in a ratio of 10: 5:1 and 50 parts waters with respect to dried biomass are used. The reactor was pressurized to 6 MPa of hydrogen pressure and standard room temperature was maintained. To eliminate residual air, the reactor is flushed with hydrogen prior to the reaction. The reactor was then heated to a desired temperature while being vigorously stirred, and it was maintained at that temperature for a specific amount of time. After the reaction, chilled water is used to cool the reactor to room temperature swiftly.

The C_5/C_6 sugars are present in the aqueous phase, which were separated by centrifugation and were then filtered to remove the solid residues and catalyst. The ZrP and Ru/C mixed catalyst exhibits high reusability. The catalyst is retrieved using centrifugation and rinsed with deionized water. A major advantage is that the catalyst mixture can be further used for the next batch without any treatment. Drying at 373 K and calcining at 673 K recover 93 wt% of the total catalyst mixture used initially. The solid residue was washed with deionized water and dried under vacuum before being used as a sample for HPLC analysis of the aqueous phase. Shylesh [43] synthesized C24–C45 cyclic compounds with excellent lubricating characteristics by self-condensation of C8–C15 alkyl methyl ketones, which were derived from biomass with the aid of Lewis acidic or basic heterogeneous catalysts. Alkylating acetone with the electrophilic alcohols created by hydrogenating C8–C16 fatty acids results in alkyl methyl ketones as product. C24–C45 cyclic alkanes are produced by base-catalyzed condensation of these alkyl methyl ketones with Mg(Al)O (calcined hydrotalcite) and subsequent HDO of the products over Pt/$NbOPO_4$. Meanwhile, aromatic lubricants are synthesized by condensation over Lewis acids, such as silica-supported Ta_2O_5, which are then hydrogenated to create cyclic alkanes. Figure 1.7 depicts the above in a flowchart.

Gallic acid is another naturally occurring organic compound that belongs to the class of phenolic acids, which is of great interest. Chemically, gallic acid is a trihydroxy benzoic acid. Its chemical formula is $C_6H_2(OH)_3COOH$. Gallic acid is a white or

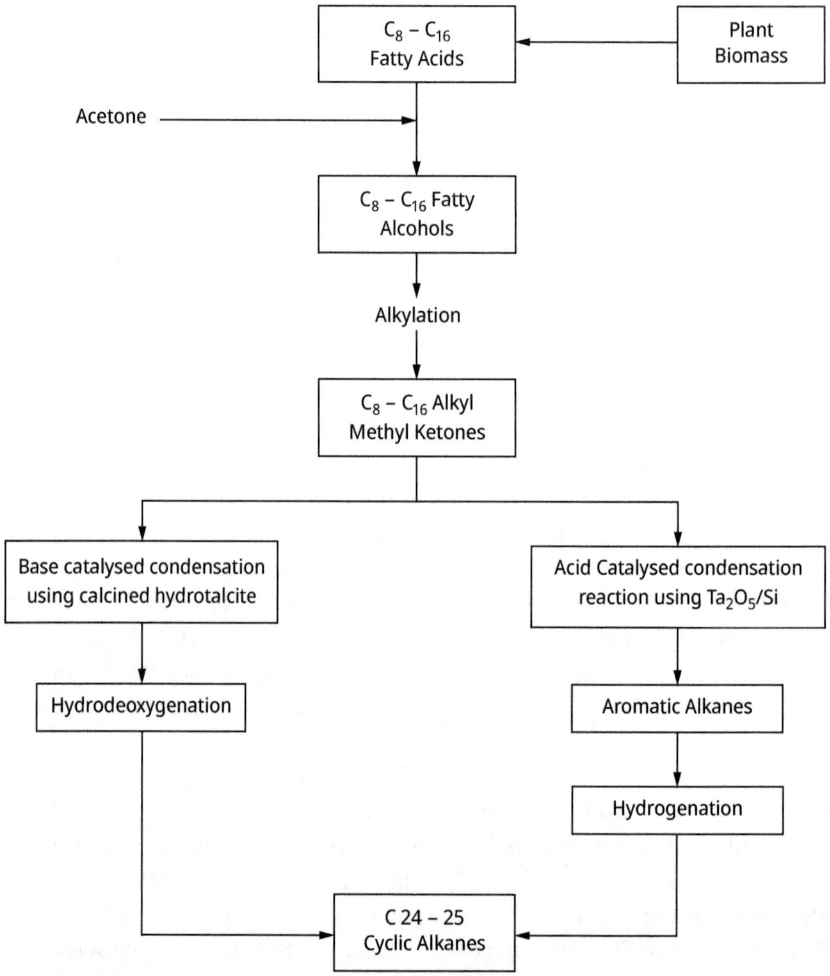

Fig. 1.7: Schematic flowchart representing the production of biolubricants from plant biomass.

pale-yellow solid that is soluble in water and ethanol, with extensive applications in food and pharmaceuticals.

Gallic acid and its ester derivatives are the primary and most crucial phenolic-type antioxidants, displaying minimal biological toxicity and great biodegradability as a class of biorenewable raw materials. Due to these materials' outstanding properties, research on their usage in many disciplines continues to grow. Due to its rigid benzene ring, three phenolic hydroxyl groups, and one carboxylic group, gallic acid can greatly enhance the physicomechanical characteristics of materials by changing the phenolic hydroxyl and carboxylic groups. The hydroxyl and carboxylic groups on gallic acid are modified to form gallate derivatives, a novel family of synthetic esters that have been shown to have excellent oxidation stability, thermal stability, and good lubricating properties, especially at elevated temperatures as shown in Fig. 1.8. Figure 1.9 shows the structures of various gallate derivatives. Amirah et al. [45] conducted a comparative study on extraction of gallic acid from different existing methods. The biomass used was the stem bark of *Jatropha curcas* and it was subjected to Soxhlet, microwave-assisted extraction (MAE), and ultrasound-assisted extraction (UAE). MAE was simple and rapid, and MAE extracts (0.5477 mg/mL) presented higher amounts of gallic acid than that of Soxhlet (0.2232 mg/mL) and UAE (0.4005 mg/mL), which were determined by HPLC.

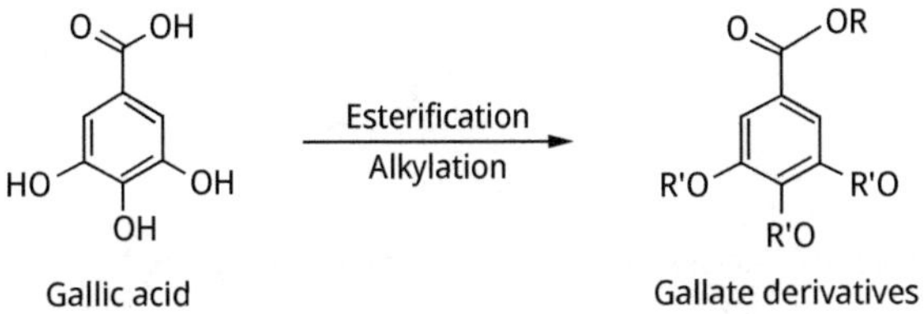

Fig. 1.8: Conversion of gallic acid to gallate derivative.

Baite et al. [46] successfully extracted 8.52 mg/g of gallic acid from the leaves of Roxburg fig (*Ficus auriculata)* using water in a sonification bath. The studies also show that the sonication bath performed 12.5 times more efficiently than probe sonicator in gallic acid extraction.

Ming-Jin et al. [47] synthesized octyl and isooctyl gallate esters and studied their performance along with methyl, propyl, and isopentyl gallate. Octyl or isooctyl alcohol, catalyst p-toluene sulfonic acid, and gallic acid were heated to 120 °C. The temperature was held constant for 4 h. The solution was shifted to a separating funnel once it was cooled to room temperature. The organic layer was separated from the aqueous layer and rinsed three times with 500 mL of distilled water to dissolve the unreacted gallic acid and p-toluene sulfonic acid. After separation, anhydrous sodium sulfate was supplied to the organic layer. It interacted with excess moisture present to form solid crystals and was easily separated by filtering. Alkyl gallate was subjected to vacuum evaporation to remove solvent and unreacted alcohol.

Distilled water was then brought into contact with the solution once cooled to room temperature. The solution was extracted three times with ethyl acetate, given brine washing, dried over Na_2SO_4, and then filtered. The crude product was refined using silica gel column chromatography using 10:1 v/v mixture of petroleum ether-ethyl acetate as eluent and excess solvent evaporated.

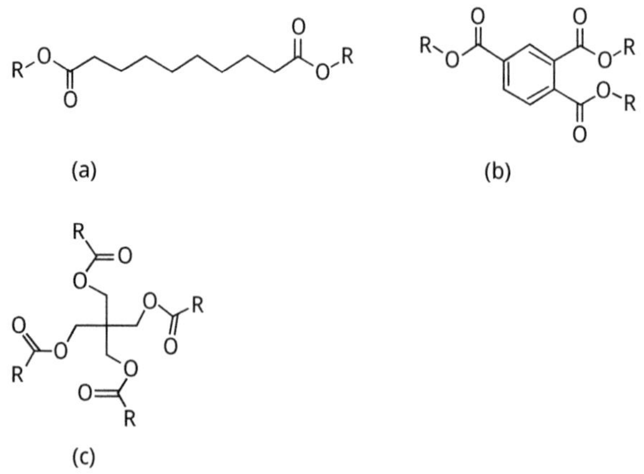

Fig. 1.9: Some gallate derivatives: (a) R = isooctyl, (b) Phe-3isoC_8, R = isooctyl, and (c) R = C_7–C_8 alkyl.

1.5 Intensification using advanced techniques

As the demands of various industries and sectors continue to grow, traditional methods alone are often inadequate to meet the challenges and complexities of modern problems. Advanced intensification techniques encompass a wide range of innovative approaches that aim to enhance efficiency, productivity, and sustainability. One compelling reason for embracing advanced intensification techniques is the pressing need to optimize resource utilization.

1.5.1 Ultrasound-assisted intensification

One of the most potent methods to intensify chemical reactions is to use ultrasonic waves. The primary process is acoustic cavitation, where gas bubbles created in a liquid media build up energy before rapidly collapsing and releasing it [48].

The surface of a liquid is affected by ultrasonic vibrations, which cause bubbles to develop and fill with liquid vapor. Bubbles oscillate as a result of repeated cycles of compression and decompression, and their volume grows when they combine with other bubbles [49]. The produced bubbles, which are alternately stretched and compressed, produce a pressure difference. During the high-pressure cycle, these bubbles then violently collapse [50]. High amount of energy is released locally when the bubbles explode, increasing the local temperature and pressure by 500 K and 100 MPa, respectively, and accelerating the overall reaction speed [49]. The cavitation event accelerates the rate of reaction by generating free radicals [51]. As a result, cavitation facilitates the entry of mass and heat into liquids. Cavitation plays an important role in speeding up chemical processes with the help of ultrasound. Numerous investigations on the manufacture of biodiesel have shown that ultrasonic irradiation is an extremely efficient method for process intensification [52, 53]. By combining two distinct phases of methanol and fatty acids in a homogeneous manner, ultrasonic irradiation enhances mass transfer. Further development of the epoxidation process for the synthesis of biolubricant uses ultrasound extensively [54, 55].

Patience et al. [56] carried out esterification reaction of 1,3-propane diol with oleic acid using ultrasound for process intensification to obtain diester. It was observed that the reaction was 2 h faster while using 500 W and 750 W ultrasound horns than conventional method involving continuous stirring. Ramachandran et al. [57] examined numerous ultrasonic-assisted transesterification reactions employing various catalysts [57], leading them to the conclusion that ultrasound lowers reaction temperature, time, and production costs for biodiesel. Ultrasound intensification is considered an eco-friendly process for several reasons. Table 1.1 lists the synthesis details of biolubricants by using ultrasound techniques.

- Reduced energy consumption: Ultrasound intensification reduces the energy requirements of chemical reactions compared to conventional methods. Ultrasonic waves can enhance mass transfer and accelerate reactions, allowing for lower.
- Operating temperatures and shorter reaction times. This energy efficiency can lead to reduced energy consumption and lower carbon emissions.
- Lower chemical usage: The reduction in chemical usage has environmental benefits, as it minimizes the amount of waste generated and the potential for pollution.
- Green solvent usage: Ultrasound can enhance the solubility and reactivity of substances in environmental-friendly solvents such as water or low-toxicity organic solvents. This reduces or eliminates the need for hazardous or volatile solvents, which are often associated with environmental and health risks.
- Alternative to harsher techniques: In industrial applications, ultrasound can replace high-temperature reactions or processes involving potentially harmful compounds, thereby reducing the impact on the environment.
- Waste reduction: Ultrasound can promote selective reactions, minimizing the formation of unwanted by-products. This reduction in waste generation can lead to a more sustainable and eco-friendly process.
- Green synthesis and cleaner manufacturing: Ultrasound has been employed in various fields, including pharmaceuticals, food processing, and wastewater treatment. It has enabled greener synthesis routes for pharmaceutical compounds, extraction of natural products, and degradation of organic pollutants. By facilitating cleaner and more efficient manufacturing processes, ultrasound intensification contributes to sustainable practices.

Tab. 1.1: Synthesis of biolubricants by using ultrasound technique.

S no.	Raw material	Co-reactants	Molar ratio of solvent: feed	Catalyst	Amount of catalyst (wt%)	Power (W)	Time (min)	Temperature (°C)	Product	Conversion	Reference
1	Oleic acid	1,3-Propane diol	Diol: oleic acid = 2:1	–	–	88.2	120	100	Diester	>99	[56]
2	Soyabean oil	Toluene, H_2O_2	H_2O_2: ethylene unsaturation = 1.5:1	Candida	4	100	300	40	Epoxide (ESO)	91	[58]
3	Rapeseed oil methyl ester	TMP	Methanol: oil = 4.63:1	–		–	180	60	TMP Triester (TMPTE)	82.21 yield	[59]
4	Stearic acid	2-Ethyl hexanol	2-Ethyl hexanol: stearic acid = 2:1	Fermase enzyme	2	80	180	50	Monoester 2- Ethylhexyl stearate	95.87	[60]

1.5.2 Other advanced techniques

An advanced method that has been investigated for its potential in the production of biolubricants is microwave irradiation. In comparison to conventional heating techniques, it has a number of benefits, including increased product yields, improved reaction rate, and selective heating.

Aguileraa et al. [61] studied the effect of microwave-assisted heating in epoxidation of oleic acid and cotton seed oil. Amberlyst 15H, a cation-exchange resin, was used as a catalyst. It was determined that the use of microwave-assisted heating resulted in a significant increase in the product yield and rate of reaction. Additionally, Barbosa et al. [62] investigated the use of microwave irradiation for solid-catalyzed transesterification, where the process took 1–2 min to complete.

Due to their limitations, these approaches require a lot of energy to sustain higher temperatures. However, it is also crucial to discover a transesterification catalyst that is more powerful. The use of supercritical solvents in biolubricant synthesis is an advanced technique that offers several advantages over traditional solvent-based processes. Supercritical solvents refer to fluids that are above their critical temperature and pressure, exhibiting properties of both liquids and gases.

A supercritical solvent is a substance that is in a state beyond its critical point, exhibiting properties of both, a liquid and a gas. The critical point is defined as the pressure and temperature at which the distinction between liquid and gas phases disappears. Table 1.2 presents the critical temperature and pressure of supercritical fluid. Solvents with low critical temperature and pressure are preferred and carbon dioxide (CO_2) is the most commonly used supercritical solvent for those reasons. However, other substances such as water, methane, ethane, propane, and various organic compounds can also be utilized as supercritical solvents depending on the application.

Tab. 1.2: Properties of selected fluids [63].

Fluid	Critical temperature (K)	Critical pressure (kPa)
Ammonia	405.55	11,649.5
Carbon dioxide	304.25	7,579.11
Ethane	305.55	5,015.588
Methanol	513.25	8,306.6
Nitrous oxide	309.75	7,435.420
Water	647.55	22,701.33
Xenon	289.85	5,996.96

Presently, the transesterification reaction was conducted without the use of catalyst by means of supercritical methanol (240 °C and 82 atm). Saka and Kusdiana [64] studied transesterification using supercritical methanol and reported that the process has several advantages. The reaction time was short compared to conventional transesterification, and also resulted in simple purification. Supercritical solvents find applications in diverse fields such as extraction, chromatography, pharmaceuticals, materials synthesis, and biolubricant production. Recently, the use of subcritical water (i.e., water between 100 and 300 °C) as a solvent has been explored to achieve hydrolysis of soybean oil without the use of catalysts [65].

1.6 Summary and conclusion

Biolubricants offer excellent lubricating properties, such as high VI and thermal stability, ensuring optimal performance and prolonged equipment life than conventional lubricants that are highly toxic and do not degrade easily. They are produced from biorenewable resources, making them more environmental friendly and sustainable. Their superior biodegradability reduces their impact on ecosystems in the case of leaks or spills.

Biolubricants are traditionally produced by chemical modifications to enhance their applicability in wider areas. KE uses renewable and off-peak energy and operates in mild conditions. But the selectivity of dimers is greatly reduced at low pressure due to the formation of non-Kolbe products. Optimum yield can be obtained at high-pressure and low-temperature conditions. The application of lipases is highly favored, due to mild reaction conditions, high selectivity, and excellent catalytic activity, thereby resulting in reduced generation of residues. This method suffers as enzymes are expensive and the reaction rates are very low. The use of biorenewable feedstock like WCO and C_5–C_6 alcohols and gallic acid can potentially replace edible oils as industrial raw material. Currently, using gallic acid as industrial feedstock for biolubricant production is not feasible due to its high cost, despite the abundance of plant biomass. Waste/used cooking oil sector is growing extensively, making it a dependable industrial raw material for biolubricant manufacturing. Advanced techniques like ultrasound irradiation and microwave-assisted techniques are also widely gaining attention. Microwave-assisted technique is simple and rapid and also shortens the reaction time. The high initial investment and operating costs associated with large-scale microwave systems have limited their adoption in many industries.

In conclusion, although eco-friendly processes often come with a higher initial cost compared to conventional methods, their adoption is expected to yield significant long-term benefits in the future. As technology advances and economies of scale are realized, the costs associated with eco-friendly processes are likely to decrease. Additionally, increased awareness of environmental sustainability and regulatory pressures are driving industries to seek greener alternatives. The potential for reduced resource consumption, minimized pollution, and enhanced corporate reputation make the investment in eco-friendly processes worthwhile. With continued research, innovation, and market demand, the cost-effectiveness of eco-friendly processes is expected to improve, leading to a more sustainable and economically viable future.

References

[1] Salih N, Salimon J. A review on eco-friendly green biolubricants from renewable and sustainable plant oil sources. Biointerface Research in Applied Chemistry. 2021;11:13303–133027.

[2] Ziogas A, Pennemann H, Kolb G. Electrochemical synthesis of tailor-made hydrocarbons from organic solvent free aqueous fatty acid mixtures in a micro flow reactor. Electrocatalysis. 2020;11:432–42.

[3] Schäfer HJ. Recent contributions of Kolbe electrolysis to organic synthesis. In: Steckhan E, ed. Topics in Current Chemistry. Electrochemistry IV. Berlin, Heidelberg: Springer, 1990, 92–143.

[4] Glasstone S, Hickling A. The hydrogen peroxide theory of electrolytic oxidation. Chemical Reviews. 1939;25:407–41.

[5] Bart JCJ, Gucciardi E, Cavallaro S. Renewable feedstocks for lubricant production. In: Bart JCJ, Gucciardi E, Cavallaro S, eds. Biolubricants. Cambridge, US: Woodhead Publishing Series in Energy, 2013, 121–248.

[6] Wang K, Tan L, inventor. Exxonmobil Research and Engineering Company, assignee. Electrochemical Synthesis to Produce Lube Stock from Renewable Feeds. United States Patent WO2014/137534A2. 2014 Sept 12.

[7] Dickinson T, Wynne Jones WFK. Mechanism of Kolbe's electrolysis. Part 1- Anode potential phenomenon. Transactions of Faraday Society. 1962;58:382–400.

[8] Vijh AK, Conway BE. Electrode kinetic aspects of the Kolbe reaction. Chemical Reviews. 1967;67:623–64.

[9] Dinh-Nguyen N. Some advantages of using Mercury cathode in the Kolbe Electro-synthesis. Acta Chemica Scandinavica. 1958;12:585–86.

[10] Sanderson JE, Levy PF, Cheng LK, Barnard GW. The effect of pressure on the product distribution in Kolbe electrolysis. Journal of the Electrochemical Society. 1983;130:1844–48.

[11] Woolford RG. The electrolysis of ω-bromocarboxylic acids. Canadian Journal of Chemistry. 1962;40:1846–50.

[12] Swann S Jr. Technique of organic chemistry. In: Weissberger A, ed. Catalytic, Photochemical and Electrolytic Reactions. 2nd ed, Vol. III. New York: Wiley-Interscience, 1956, 385–95.

[13] Reeves CJ, Menezes PL. Advancements in eco-friendly lubricants for tribological applications: Past, present, and future. In: Davim JP, ed. Ecotribology Research Developments. New York, US: Springer Cham, 2016, 41–61.

[14] Calvin K, Kimberly B, Brant A. Biolubricants through renewable hydrocarbons: A perspective for new opportunities. Renewable and Sustainable Energy Reviews. 2019;113:109261.

[15] Malevich D, Gibson GTT. inventor. Advonex International Corp., assignee. Production of hydrocarbons from plant oil and animal fat. United States Patent US 9,611,554 B2. 2017 April 4.

[16] Börner A, Franke R. Hydroformylation: Fundamentals, Processes, and Applications in Organic Synthesis. Vol. II. Weinheim, Germany: John Wiley & Sons, 2016.

[17] Varone A, Ferrari M. Power to liquid and power to gas: An option for the German Energiewende. Renewable and Sustainable Energy Reviews. 2015;45:207–18.

[18] Bolina CAI, Gomes ABR, Mendes AA. Production from several oleaginous feedstocks using lipases as catalysts: Current scenario and future perspectives. BioEnergy Research. 2021;14:1039–57.
[19] Thangaraj B, Solomon PR. Immobilization of lipases–a review. Part I: Enzyme immobilization. chemical and Biochemical Engineering Reviews. 2019;6:157–66.
[20] Angajala G, Pavan P, Subashini R. Lipases: An overview of its current challenges and prospectives in the revolution of biocatalysis. Biocatalysis, Agricultural Biotechnology. 2016;7:257–70.
[21] Ferreira-Leitão VS, Cammarota MC, Aguieiras ECG, et al. The protagonism of biocatalysis in green chemistry and its environmental benefits. Catalysts. 2017;7:9.
[22] Robinson KP. Enzymes: Principles and biotechnological applications. Essays Biochemistry. 2015;59:1–41.
[23] Singh RS, Singh T, Pandey A. Microbial enzymes – an overview. In: Singh RS, Singhania RR, Pandey A, Larroche C, eds. Advances in Enzyme Technology. Biomass, Biofuels, Biochemicals. Cambridge, US: Elsevier, 2019.
[24] Cavalcanti da Silva JA. Biolubricant production catalyzed by enzymes. In: Sharma KB, Biresaw G, eds. Environmentally Friendly and Biobased Lubricants. Boca Raton: CRC Press, 2016.
[25] Jaeger KE, Reetz MT. Microbial lipases form versatile tools for biotechnology. Trends in Biotechnology. 1998;16:396–403.
[26] Castro HF, Mendes AA, Santos JC, Aguiar CL. Modification of oils and fats by biotransformation. Química Nova. 2004;27:146–56.
[27] Malcata FX, Reyes HR, Garcia HS, et al. Immobilized lipase reactors for modification of fats and oils-A review. Journal of American Oil Chemists' Society. 1990;67:890–910.
[28] Musa U, Mohammed IA, Sadiq MM, Aberuagba F, Olurinde AO, Obamina R Synthesis and characterization of trimethylolpropane-based biolubricants from castor oil. In: Proceedings of the 45th Annual Conference of NSChE. Warri, Nigeria, 2015 Nov 5–7, 248–54.
[29] Cavalcanti da Silva JA, Soares VF, Fernandez- Lafuente R, Habert AC, Freire DMG. Enzymaticproductionandcharacterizationofpotentialbiolubricantsfromcastorbeanbiodiesel. Journal of Molecular Catalysis B: Enzymatic. 2015;122:323–29.
[30] Arslan A, Rancke-Madsen A, Brask J. Enzymatic synthesis of estolides from castor oil. Catalysts. 2020;10:835–44.
[31] Liu J, Liu A, Hu Y. Enzymatic dimerization in the biosynthetic pathway of microbial natural products. Natural Products Reports. 2021;38:1469.
[32] Laane C, Cross GA, Beindorff C, Haase D, Kloosterman J. (Bio)dimerization of unsaturated fatty acids. Recueil Des Travaux Chimiques Des Pays-Bas. 1991;110:195–99.
[33] Gui MM, Lee KT, Bhatia S. Feasibility of edible oil vs. non-edible oils vs. Waste edible oil as biodiesel feedstock. Energy. 2008;33:1646–53.
[34] Wan N, Mastura B, Ramadhansyah PJ, Mohd RH, Mohamed A, Norzita N, et al. The potential of waste cooking oil as bio-asphalt for alternative binder – An overview. Journal of Tecknologi. 2016;78:111–16.
[35] Moghaddam TB, Mohamed RK, Abdelaziz M. A review on fatigue and rutting performance of asphalt mixes. Scientific Research and Essays. 2011;6:670–82.
[36] Awogbemi O, Von Kallon DV, Aigbodion VS, Panda S. Advances in biotechnological applications of waste cooking oil. Case Studies in Chemical and Environmental Engineering. 2021;4.
[37] Singhabhandhu A, Tezuka T. The waste-to-energy framework for integrated multi-waste utilization: Waste cooking oil, waste lubricating oil, and waste plastics. Energy. 2010;35:2544–51.
[38] Mannu A, Garroni S, Porras JI, Mele A. Available technologies and materials for waste cooking oil recycling. Processes. 2020;8:366–79.
[39] Chowdhury A, Mitra D, Biswas D. Biolubricant synthesis from waste cooking oil via enzymatic hydrolysis followed by chemical esterification. Journal of Chemical Technology & Biotechnology. 2012;88:139–44.

[40] Jahromi H, Adhikari S, Roy P, Shelley M, Hassani E, Oh TS. Synthesis of novel biolubricants from waste cooking oil and cyclic oxygenates through an integrated catalytic process. ACS Sustainable Chemistry and Engineering. 2021;9:13424–37.

[41] Zhang W, Ji H, Song Y, Ma S, Xiong W, Chen C, et al. Green preparation of branched biolubricant by chemically modifying waste cooking oil with lipase and ionic liquid. Journal of Clean Products. 2020;274:122918.

[42] Zhang S, Maréchal F, Gassner M. Process and integration of fuel ethanol production from lignocellulosic biomass based on double acid hydrolysis. Energy Fuels. 2009;23:1759–65.

[43] Shylesh S, Gokhale A, Ho C, Bell A. Novel strategies for the production of fuels, lubricants and chemicals from biomass. Accounts of Chemical Research. 2017;50:2589–97.

[44] Lui Q, Zhang T, Liao Y, Cai C, Tan J, Wang T, et al. Production of C_5/C_6 sugar alcohols by hydrolytic hydrogenation of raw lignocellulosic biomass over Zr based solid acids combined with Ru/C. Sustainable Chemistry and Engineering. 2017;5:5940–50.

[45] Amirah, Reddy P, Khan M. Comparison of extraction techniques of gallic acid from Stem Bark of *Jatropha curcas*. Journal of Applied Sciences. 2012;12:1106–11.

[46] Baite TN, Mandal B, Purkait MK. Ultrasound assisted extraction of gallic acid from Ficus auriculata leaves using green solvent. Food and Bioproducts Processing. 2021;128:1–11.

[47] Ming-Jin F, Zhang C, Wen P, Sun W, Dong R, Yang DS, et al. High-performance lubricant base stocks from biorenewable gallic acid: Systematically study on their physicochemical and tribological. Industrial & Engineering Chemistry Research. 2017;56:9513–23.

[48] Fang Z, Smith RL, Qi X. Production of Biofuels and Chemicals with Ultrasound. Dordrecht, the Netherlands: Springer, 2015.

[49] Hajinezhad A, Abedi S, Ghobadian B, Noorollahi Y. Biodiesel production from Norouzak (Salvia lerifolia) seeds as an indigenous source of bio fuel in Iran using ultrasound. Energy Conversion Management. 2015;99:132–40.

[50] Leong TSH, Martin GJO, Ashokkumar M. Ultrasonic encapsulation – A review. Ultrasonics Sonochemistry. 2017;35:605–14.

[51] Jalali MR, Sobati MA. Intensification of oxidative desulfurization of gas oil by ultrasound irradiation: Optimization using Box–Behnken design (BBD). Applied Thermal Engineering. 2017;111:1158–70.

[52] Kashyap SS, Gogate PR, Joshi SM. Ultrasound assisted intensified production of biodiesel from sustainable source as karanja oil using interesterification based on heterogeneous catalyst (γ-alumina). Chemical Engineering and Processing- Process Intensification. 2019;136:11–16.

[53] Fayyazi E, Ghobadian B, Najafi G, Hosseinzadeh B, Mamat R, Hosseinzadeh J. An ultrasound-assisted system for the optimization of biodiesel production from chicken fat oil using a genetic algorithm and response surface methodology. Ultrasonics Sonochemistry. 2015;26:312–20.

[54] Chavan AP, Gogate PR. Ultrasound assisted synthesis of epoxidized sunflower oil and application as plasticizer. Journal of Industrial and Engineering Chemistry. 2015;21:842–50.

[55] Chavan VP, Patwardhan AV, Gogate PR. Intensification of epoxidation of soybean oil using sonochemical reactors. Chemical Engineering and Processing- Process Intensification. 2012;54:22–28.

[56] Patience N, Galli F, Rigamonti M, Schieppati D, Boffito D. Ultrasonic intensification to produce diester biolubricants. Industrial & Engineering Chemistry Research. 2019;58:7957–63.

[57] Ramachandran K, Suganya T, Nagendra Gandhi N, Renganathan S. Recent developments for biodiesel production by Ultrasonic assist transesterification using different heterogeneous catalyst: A review. Renewable Sustainable Energy Reviews. 2013;22:410.

[58] Bhalerao MS, Kulkarni VM, Patwardhan AV. Ultrasound-assisted chemoenzymatic epoxidation of soybean oil by using lipase as biocatalyst. Ultrasonics Sonochemistry. 2018;40:912–20.

[59] Almasi S, Ghobadian B, Najafi G, Soufi MD. A novel approach for bio- lubricant production from rapeseed oil- based biodiesel using ultrasound irradiation: Multi- objective optimization. Sustainable Energy Technologies and Assessments. 2021;43:100960.

[60] Gawas S, Rathod V. Ultrasound assisted green synthesis of 2-Ethylhexyl stearate: A cosmetic biolubricant. Journal of Oleo Science. 2020;69:1043–49.

[61] Aguileraa AF, Tolvanena P, Herediaa S, Muñoza MG, Samsonb T. Ogerb A et al. Epoxidation of fatty acids and vegetable oils assisted by microwaves catalyzed by a cation exchange resin. Industrial and Engineering Chemistry Research. 2018;57:3876–86.

[62] Barbosa SL, Dabdoub MJ, Hurtado GR, Klein SI, Baroni AC, Cunha C. Solvent free esterification reactions using Lewis acids in solid phase catalysis. Applied Catalysis A: General. 2006;313:146–50.

[63] Hitchen SM, Dean JR. Properties of supercritical fluids. In: Dean JR, ed. Applications of Supercritical Fluids in Industrial Analysis. Dordrecht, Netherlands: Springer, 1993.

[64] Almeida L, Corazza ML, Sassaki GL, Voll FAP. Experimental study and kinetic modelling of waste frying soybean oil hydrolysis in subcritical water. Reaction Kinetics, Mechanism and Catalysis. 2017;121:439–52.

[65] Carr AG, Mammucari R, Foster NR. A review of subcritical water as a solvent and its utilisation for the processing of hydrophobic organic compounds. Chemical Engineering Journal. 2011;172:1–17.

Sergio Nogales-Delgado, Juan Félix González, José María Encinar

2 Biolubricants based on vegetable oils as a suitable replacement for petrol-based lubricants

2.1 Introduction

2.1.1 Global scenario and the role of biolubricants

It is well known that there is an increasing concern, from local to international agencies and from individuals to global populations, about two apparently different subjects like environmental conservation (and what it implies, like green chemistry, sustainability, circular economy, etc.) and economic future (highly related to geopolitical factors where energy and materials sources play a vital role) [1].

A prime example is the UN and its Sustainable Development Goals, where the compliance with sustainable practices could contribute to the sustainable economic growth of developing areas or countries, trying to be as much economic independent as possible [2]. In this regard, sustainability and economics have been as related as ever, using green technologies to combat inflation or price volatility mainly related to fuel market or geopolitical aspects, which is already showing changes in the current and global energy scenario [3, 4].

Although many of the national and international energy policies are focused on energy production [5], if we consider UN strategies, some Sustainable Development Goals defined by the UN could correspond with practices related to biolubricant production and its replacement for traditional lubricants, as observed in Tab. 2.1. In this table, some Sustainable Development Goals and their corresponding targets having to do with biolubricant production are included.

Thus, as observed in this table, the production and use of biolubricants seem to be an ideal practice for green chemistry and sustainable practices, complying with many of the Sustainable Development Goals and their corresponding targets.

Indeed, the implementation of technologies based on vegetable oils to produce biolubricants (among other products, as discussed later in this chapter) would be perfect to reuse wastes, obtain biodegradable products with a low environmental impact, or contribute to the sustainable economic growth of developing countries or areas, as different crops easily adapted to extreme climate conditions could be used for this purpose.

Additionally, the activity of petrol-based refineries, focused on products like diesel and lubricants (main subject matter in this chapter), as well as the use of these products, have been proven to be extremely pollutant, contributing to undesirable environmental effects such as greenhouse gas effect, acid rain, or water pollution,

https://doi.org/10.1515/9783111145136-002

Tab. 2.1: Examples of some Sustainable Development Goals (SDGs) and their relationship to biolubricant production.

SDGs	Role of biolubricants
7. Affordable and Clean Energy	Some chemical routes, such as transesterification of fatty acids, could produce biofuels like fatty acid methyl esters, mainly used as biodiesel, contributing to the share of renewable energy in the global energy mix.
8. Decent Work and Economic Growth	Biolubricants can be obtained from local vegetable oil crops, contributing to the per capita economic growth in accordance with national circumstances. Also, the role of biolubricants in biorefineries could assure economic productivity diversification.
9 and 11. Industry, Innovation and Infrastructure; Sustainable Cities and Communities	The use of biorefineries based on vegetable oils could promote inclusive and sustainable industrialization, and the use of biodiesel and biolubricants could provide access to safe, affordable, accessible, and sustainable transport systems, also contributing to air quality preservation, among others.
12. Responsible Consumption and Production	Biolubricant production and consumption is sustainable, making the management of natural resources more efficient. Also, some food wastes like used frying oil could be recycled to produce biolubricants, rationalizing fossil products.
13. Climate Action	Biolubricant and biodiesel production and use could reduce different climate-related hazards such as acid rain or greenhouse gas emission, which is important to integrate different climate change measures at national and international levels.
14 and 15. Life Below Water; Life on Land	As biolubricants are more biodegradable compared to traditional lubricants, its environmental impact in case of spill in water or land would be reduced.

among others [6]. That is the reason why, from many different points of view such as social, political, entrepreneurial, or scientific, the search for alternatives to these harmful activities has become vital in the last two decades [7–9].

This way, there are some interesting advantages related to the production or use of biolubricants, as observed in the following points [8, 10, 11]:

- Biolubricant production can take place in a biorefinery context, with the subsequent advantages related to these kinds of facilities, as explained in further detail in the following sections.
- Its production implies zero-net CO_2 emissions, even though some biofuels like biodiesel are obtained.

- They can be obtained from different raw materials, such as vegetable oils from many sources (such as crops or wastes) or microalgae, which can be easily adapted to different countries or regions.
- This wide range of raw materials make biolubricant production a process easily implemented or adapted to many different countries or regions, especially in developing areas.
- Their tribological behavior, if properly produced, is similar to their respective equivalent based on petroleum.
- Indeed, some properties are improved when biolubricants are used instead of traditional lubricants, such as an increase in flash and combustion points, which are generally higher for the former, implying safety during storage of these products, avoiding combustion at high temperatures.
- Taking into account that they come from natural sources, their molecular structure is usually similar to those found in their corresponding raw material, which makes them more biodegradable and easily degraded through biochemical routes carried out by microorganisms. Consequently, their environmental impact in the case of accidental spillage would be less harmful and durable in terrestrial and aquatic ecosystems compared to other products like oil or traditional lubricants.

2.1.2 Scientific interest in this field

Consequently, there is an increasing interest in vegetable oils as a starting point for biolubricant production by the scientific community, as observed in Fig. 2.1, for published articles in impact factor journals. Thus, there was a considerable increase in articles from 2010, and this trend seems to be exponential in the past 5 years. This fact is a reflection of a real and reliable interest in the replacement of traditional lubricants for these bio-based products.

Additionally, this increasing interest comes from different and multidisciplinary fields, as observed in Fig. 2.2. This fact points out the importance of biolubricant production, and the required involvement of different disciplines to implement this technology at an industrial level with a high level of industrial development.

This way, considering Fig. 2.2, engineering and chemical engineering are the major fields where scientific articles were published, which can be explained by the fact that the implementation of a biorefinery is one of the key points to make this technology feasible and efficient compared to traditional refineries based on petrol. With 33% of total articles, chemistry, energy, and materials science are possibly focused on the foundations of chemical reactions taking place (like transesterification and epoxidation), energy requirements, or energy products obtained during the process (such as biodiesel in double transesterification) and properties for lubrication, respectively.

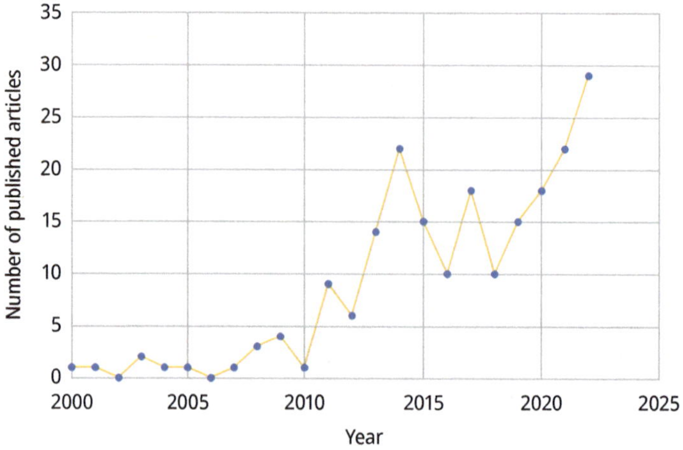

Fig. 2.1: Evolution of published articles in the last 25 years about biolubricants from vegetable oils [12].

Also, other subjects that have a reduced impact but are also interesting are: environmental sciences, where the study of environmental factors such as biodegradability of biolubricants could be the key point in these kinds of research works; agricultural and biological sciences, focused on the possibility of different crops for the production of biolubricants, paying attention to the yield of the process and quality control; biochemistry and genetics, which could be focused on the study of genetically modified crops to improve the quality of vegetable oils used in biolubricant production, implying the im-

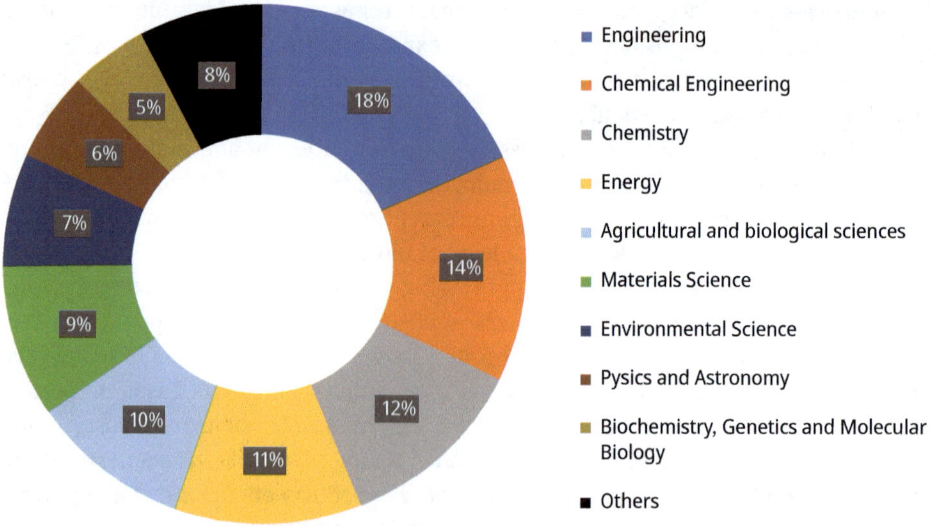

Fig. 2.2: Article distribution according to different scientific fields [12].

provement of the subsequent products obtained during the process, like an increase in oxidative stability of biodiesel and biolubricant.

On the other hand, it is interesting to consider a worldwide distribution of scientific publications dealing with this subject. For this purpose, Fig. 2.3 shows the main countries where articles dealing with biolubricants based on vegetable oils come from. Interestingly enough, China and the United States (scientific world powers) are not included in the top five of this list, whereas Malaysia, India, Spain, Brazil, and Iraq were the countries with most published articles about this subject. Consequently, and taking into account that safflower (especially in India) and soybean are very popular in Asia, it is no wonder that the number of research works based on these vegetable oils is considerable.

It is noticeable that there is a relatively homogeneous distribution of research carried out about this subject, with a country from Africa (Nigeria) in this top ten list. This point is quite encouraging, as it means that the implementation of this technology is considered worldwide from a scientific and technical perspective, pointing out the good adaptation of raw materials suitable for biolubricant production. Additionally, the interest of some African countries is also positive and significant, as it is evidence that serious attempts to implement biorefineries based on vegetable oils are considered in developing or emerging countries where there are many problems related to energy dependence.

In any case, the potential of biolubricant production can be extended to many other countries, especially where there is a long-established agricultural practice

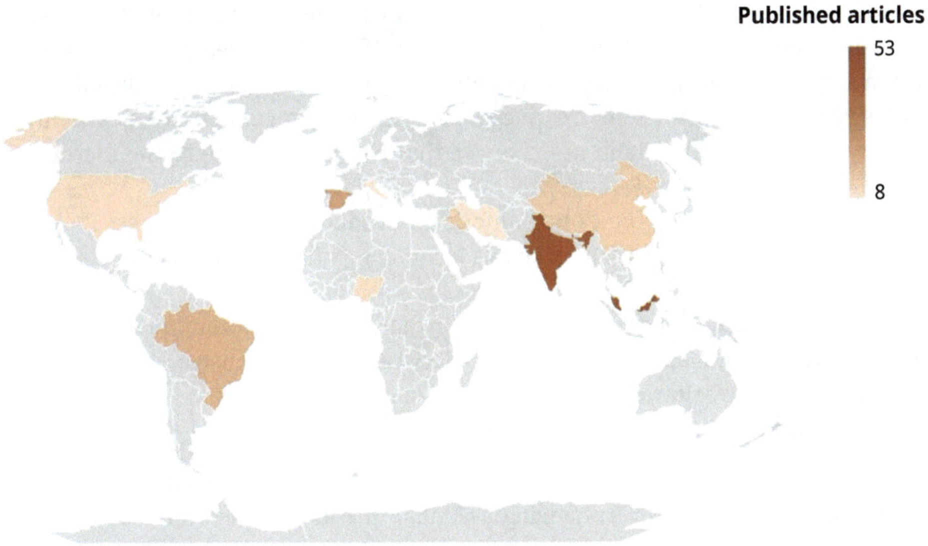

Fig. 2.3: Top 10 countries with published articles about biolubricants from vegetable oils (with Bing technology) [12].

without any access to direct petroleum sources. Therefore, it would not be surprising if the global outlook observed in Fig. 2.3 changes over time.

In conclusion, it could be said that the use of vegetable oils as a raw material to produce biolubricants is a promising field, where different scientific disciplines could take part to improve the implementation of this technology to make lubrication in industry (and society in general) sustainable. However, the origin of these raw materials could be controversial, as it happened in other similar products like biodiesel, which will be discussed in the following section.

2.2 Raw materials: can biolubricant production compromise food crops?

According to the previous section, plenty of studies about the implementation of vegetable oils as biodiesel or biolubricant source have been carried out, which implies the use of a wide range of vegetable oils, both food and nonfood vegetable oils. For instance, castor, jatropha, palm, rapeseed, safflower, soybean, or sunflower, among others, have been considered to produce biodiesel or biolubricants [13].

Regardless of the use of these crops, it is clear that its intensive promotion in new green policies might be controversial, as apart from the obvious advantages related to the replacement of petrol-based refineries to obtain more pollutant products, there are some issues that could challenge the implementation of biorefineries based on vegetable oils to a larger extent, especially concerning environmental and industrial issues.

For instance, regarding biodiesel production (a highly related process as we will see in further sections) at a large scale, some ethical issues could arise, like oil spill in water after biodiesel washing treatment, food vs fuel competition, the possibility of deforestation, and other land-use changes that could contribute to the increase in greenhouse gas emissions [14].

Nonetheless, when it comes to biolubricant production, some thoughts should be taken into consideration, which make this technology suitable and sustainable:

– Biolubricant requirements are much lower than biodiesel consumption in automotive engines, which makes its production selective and less intensive from an agricultural point of view. However, the industrial activity might increase the demand for lubricants in machines, engines, and facilities.

– Biolubricant production from vegetable oils could contribute to a diversification of a biorefinery's performance, changing and adapting its production to specific products depending on demand, which can change on account of dry seasons, flood, or even geopolitical factors that might have influence on oil crop exports such as sunflower from warring countries like Ukraine or Russia.

– As observed in biodiesel production, some crops (not necessarily non-food) could be easily adapted to poor soil or extreme climate conditions, not conflicting with food crops [15]. What is more, some crops like cardoon, rapeseed, or safflower could take part in crop rotation, as these plants present a tap root that tends to deepen enough in soils, so that other crops like wheat can take advantage of superficial soil when it is recovered. Specifically, the use of these three vegetable oils has been proven to be effective for biolubricant production, obtaining high-quality products with a desirable viscosity value and high flash and combustion points. Also, depending on their fatty acid content, their oxidative stability could vary, generally requiring certain amounts of antioxidants (as we will see later) [16–20].

– Nevertheless, vegetable oils for biolubricant production do not always necessarily imply the direct use of crops. Thinking about used frying oils, which come directly from vegetable oils and are a waste with difficult management, it could be an interesting alternative for biolubricant production, making the valorization of such problematic waste (let's not forget its high pollutant capacity, especially in aquatic ecosystems) possible. Thus, homes, restaurants, and food industry might be an interesting source of this waste. Moreover, if we take into account the growing demand for vegetable oils for frying or sautening purposes, as shown in Fig. 2.4 in the case of the United States (where people using oil for frying or sautening has increased by approximately 15% in the last decade), a growing concern is expected about used frying oils in the future, making the possibility of its use in a biorefinery necessary and feasible, as the nature of this oil is not excessively different from an unused oil after a suitable pretreatment. Indeed, waste frying oil has already been used as an interesting source to produce biodiesel (i.e., fatty acid methyl esters, which is usually one of the intermediate products during some kinds of biolubricant production), with successful results [21–25]. Thus, considering these products equivalent to biolubricants (in many occasions biolubricants are complex esters, whereas biodiesel is composed of fatty acid methyl esters), it is not surprising that some studies have dealt with biolubricant production from used frying oils, obtaining satisfactory results and high-quality products (high conversion yields at relatively mild reaction conditions and good properties like high-viscosity values and flash and combustion points, although oxidative stability did not exceed 4 h in most cases) [26, 27].

To sum up, biolubricant production in a biorefinery context presents many environmental advantages, such as an easy implementation in biodiesel facilities and biodegradability (as will be seen in the following section), that offset the above-mentioned concerns or issues (which are equally lessened compared to biodiesel production, for example). In addition, taking into account the increasing production of some wastes like used frying oil, the use of this waste to produce biolubricants could justify the implementation of this technology in the medium or long term.

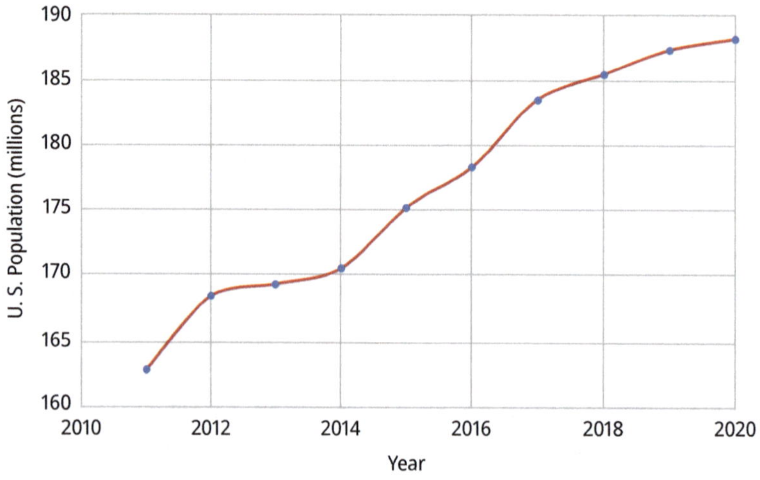

Fig. 2.4: The US population using vegetable oils for pan frying or sautening [28].

2.3 Biolubricant production: properties "on demand" in a biorefinery context

2.3.1 Different chemical routes to produce biolubricants

Regarding biolubricant production, there are different chemical routes to obtain compounds that can be used in lubrication processes, such as epoxidation and transesterification, among others. Indeed, there is not only a limited way to produce biolubricants, as many different substances or products can be used for this purpose. Thus, according to Tab. 2.2, the main chemical routes for biolubricant production are included [13, 29, 30].

Tab. 2.2: Main chemical routes to produce biolubricants.

Chemical route	Biolubricant characteristics
Estolide formation	Improved lubricity and oxidative stability, pour point decrease
Epoxidation, ring opening, and acetylation	Better oxidative stability and lubricity, decrease in viscosity index
Transesterification	High thermo-oxidative stability, pour point decrease

This chapter focuses on the latter chemical route, where triglycerides react with methanol or ethanol to produce fatty acid methyl esters (or fatty acid ethyl esters), to continue with a second transesterification process with superior alcohol to obtain biolubricants. In Fig. 2.5 the main steps and products obtained in this double transesterification are included.

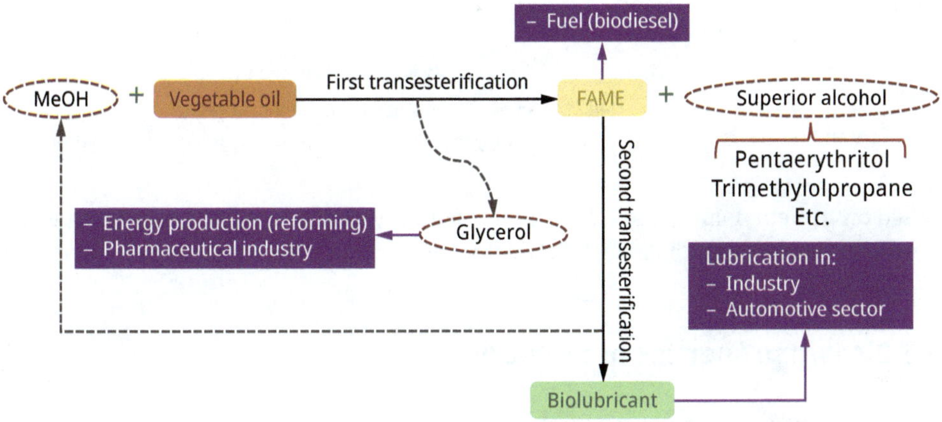

Fig. 2.5: Biolubricant production through double transesterification in a biorefinery context.

In dashed circles, reagents (or products) are related to the first and second transesterification. In purple boxes, the main uses of different products are obtained.

This way, and depending on the kind of alcohol used in the above-mentioned biorefinery, a wide variety of products can be obtained, basically changing the molecular complexity of the alcohol. For instance, by using methanol, biodiesel is obtained, whereas the use of other complex alcohols like those observed in Fig. 2.6 produces biolubricants. Essentially, biodiesel and biolubricants differ in terms of viscosity (biodiesel has viscosity values between 4 and 5 cSt, whereas in the case of biolubricants, these values are considerably higher), which is directly related to the molecular structure obtained.

Regarding this fact, the molecular structure of fatty acid methyl esters, which is less complex than those polyesters obtained from alcohols like neopentylglycol (2,2-dimethylpropane-1,3-diol), trimethylolpropane (2-ethyl-2-hydroxymethyl-1,3-propanediol), or pentaerythritol (2,2-Bis(hydroxymethyl)-1,3-propanediol), will imply a product with low viscosity compared to the latter, where there are more intermolecular interactions such as Van der Waals forces or hydrogen bonds.

In that sense, different factors can influence the viscosity of a specific biolubricant, such as branching, the presence of special functional groups that could contribute to the increase in hydrogen bonds (with the subsequent increase in viscosity), or the ability to generate di-, tri-, or tetraesters. This fact offers a great opportunity to produce biolubricants "on demand," as products with a wide range of viscosity values can be obtained, being useful for specific lubrication purposes. Thus, biolubricants

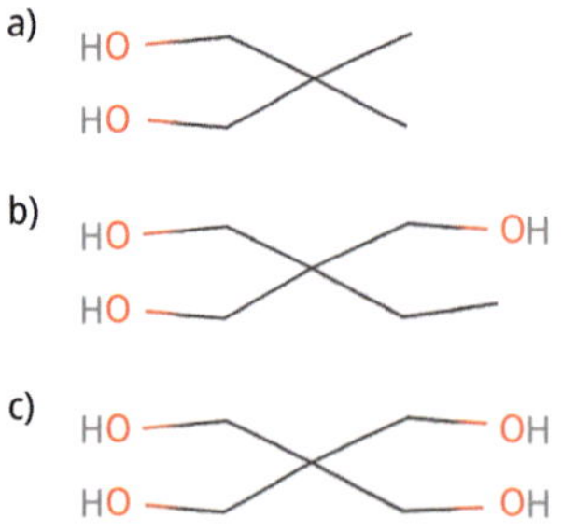

Fig. 2.6: Some alcohols used in biolubricant production: (a) neopentylglycol; (b) trimethylolpropane; and (c) pentaerythritol.

based on trimethylolpropane (TMP) [30] or pentaerythritol (PE) [18, 31] might offer different properties when it comes to viscosity and oxidative stability values.

2.3.2 Main properties of biolubricants

As a consequence of the above, and depending on the process and kind of superior alcohol selected for the process, the main properties that might be influenced by double transesterification are the following:

- **Viscosity:** This property might be influenced by the complexity of the molecular structure. As it will be explained in oxidation processes, the more complex a molecule is, the more interactions like hydrogen bonds or van der Waals forces will take place, implying at a macroscopic level a higher resistance to flow, that is, viscosity. Also, the presence of functional groups could contribute to the increase in this property.
- **Viscosity index:** In general, biolubricants obtained through double transesterification present a high viscosity index, which is a very interesting property that implies lower changes in viscosity with temperature. Consequently, these products will present stable properties if the room or ambient temperature is changed.
- **Oxidative stability:** As in the case of biodiesel samples, oxidative stability is highly influenced by the molecular structure and, especially, the presence of double bonds [32].

In any case, double transesterification presents interesting conditions to be considered in a biorefinery context (as explained in the following section), especially concerning its versatility when producing interesting products with specific properties depending on operating conditions.

2.3.3 Biorefinery concept in double transesterification processes to produce biolubricants

Regarding double transesterification from vegetable oils, this chemical route could be an optimal way to implement industry in a biorefinery context, as it presents many characteristics or advantages like the following [33, 34]:

- They are based on sustainable production, as vegetable oils are natural raw materials, including frying oil, waste with a difficult management.
- The process includes the generation of a wide range of products, such as fatty acid methyl esters (used as biodiesel), fatty acid esters (used as biolubricants), glycerol (an interesting energy source or pharmaceutical product, depending on its purity), methanol (a product generated in the first transesterification which can be reused in the first transesterification). Apart from that, other wastes related to vegetable oil generation can be equally valued, such as agricultural wastes or cake (after mechanical press for oil extraction).
- Considering the above, the theoretical atom efficiency of the process can be high (equivalent to 100% in this case). The role of efficiency in this context is vital, as high conversions (exceeding 90% or 95%) of triglycerides in the first transesterification and fatty acid methyl esters in the second transesterification will provide a real efficiency process very close to the theoretical value.
- The above-mentioned wide range of products obtained can make this technology adaptable to different demands, making the implementation of this technology feasible and suitable to the current and changing energy and product market, avoiding economic dependency due to geopolitical factors, among others.
- The products obtained are highly biodegradable, as they are obtained from natural and biodegradable sources, implying fewer problems in case of accidental spills in natural habitats.

Indeed, biodiesel production has been widely studied as an important part of biorefineries based on vegetable oils, coupled with other technologies such as biomethane production [35, 36]. It proves the high versatility of transesterification of triglycerides with methanol, which can be equally coupled to a second transesterification to produce biolubricants in a biorefinery context [37]. In any case, there are many advantages related to the use of biolubricants in a biorefinery context, but there are some advantages that can constitute a challenge for further research in this field.

2.4 Some challenges: are biolubricants stable enough?

Many advantages related to biodiesel and biolubricants, compared to their equivalent petrol-based products, have already been mentioned in this chapter. However, there are some disadvantages are a great opportunity for scientific research to make biolubricant production suitable at an industrial scale. With this regard, one of the main challenges is the usually low oxidative stability of biolubricants. Indeed, it is a typical problem in biodiesel samples (an intermediate product in double transesterification processes to obtain biolubricants), which share similar molecular structure, where auto-oxidation processes during storage are a major problem.

In this case, the presence of unsaturations seems to be the starting point for auto-oxidation (initiation), with free radical generation and the subsequent propagation and termination processes. Moreover, the combination of free radicals can imply free fatty acid generation (which increases acid number in samples, being detrimental for industrial facilities) or polymerization (causing an increase in viscosity as intermolecular interactions like Van der Waals forces take place, increasing the resistance to flow of the fluid) [38–40]. Indeed, a considerable increase in viscosity and acid number in biolubricants (without any antioxidant addition) during extreme oxidation conditions has been observed, with up to 1,000% and 700% increase in viscosity and acid number, respectively, which made these products unable to be used for their original purpose in industry [18, 31].

2.4.1 The role of oxidative stability and its assessment

Consequently, oxidative stability, which can be easily measured through the Rancimat method [38], would give an approximate idea about the performance of a biolubricant during storage or oxidation processes. In short, as included in Fig. 2.7, a biolubricant sample (3 g) is bubbled with air at 110 °C, and the degradation components released by the biolubricant (such as aldehydes, carboxylic acids, etc.), included in air, are dissolved in distilled water, whose conductivity is monitored by using a conductivity meter.

To determine the oxidative stability (or induction point) of biolubricants, it is important to consider the conductivity evolution of water. Thus, according to Fig. 2.7, there are two clear stages during this experience: a stable one, where conductivity does not increase much, and an unstable one, where conductivity considerably increases due to the abrupt release by biolubricants and the subsequent dilution in distilled water. If these stages are extended by a line, the time at which the cross point obtained appears is the so-called induction point, which is usually expressed in hours or minutes, depending on the stability of the sample.

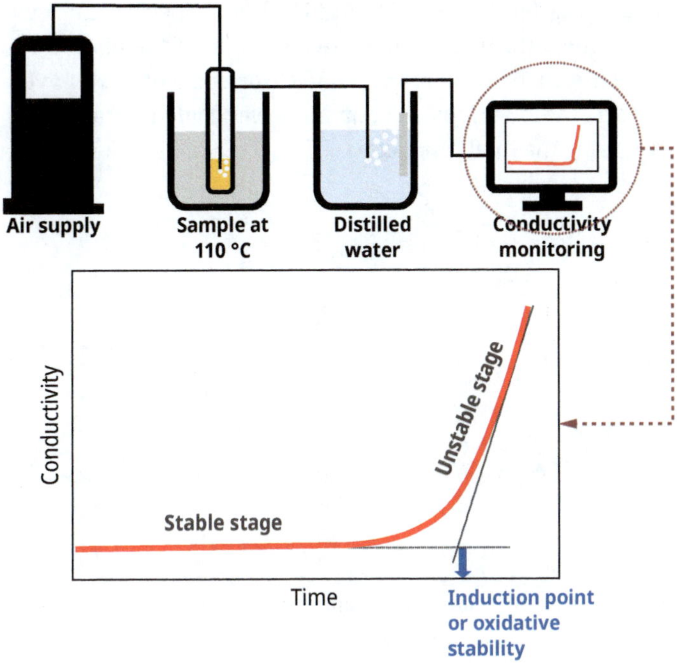

Fig. 2.7: Rancimat method and oxidative stability determination.

As stated above, the presence of unsaturations could be a problem in this respect, especially when these double bonds are not altered during the chemical reactions taking place in biolubricant production (as in the case of double transesterification).

2.4.2 Key factors affecting oxidative stability

Thus, in biolubricants obtained through epoxidation processes, this problem could be attenuated, as long as the product offers a good performance for a specific lubricating purpose. However, for double transesterification, the same concern is observed for biodiesel and biolubricants. Indeed, some key factors can help us understand the oxidative stability of a sample, including different solutions to increase this property, as follows:

– **Fatty acid composition**: The nature of the raw material is vital to understand the oxidative stability of biodiesel and biolubricant, as they share similar characteristics when it comes to molecular structure (including double bonds). With this regard, the fatty acid composition of vegetable oils plays an important role in oxidative stability, increasing (or decreasing) this value depending on the majority fatty acids in the sample. As observed in Fig. 2.8, different vegetable oils can present very different fatty acid profiles, which can imply a different behavior in biolubricant performance depending on fatty acid composition, affecting the oxidative stability of the oil and the subsequent products. In that sense, high percentages of linoleic or linolenic acids in vegetable oils such as cardoon, corn, safflower, soybean, or sunflower could indicate lower oxidative stability values regarding the final products (i.e., biodiesel and biolubricant).

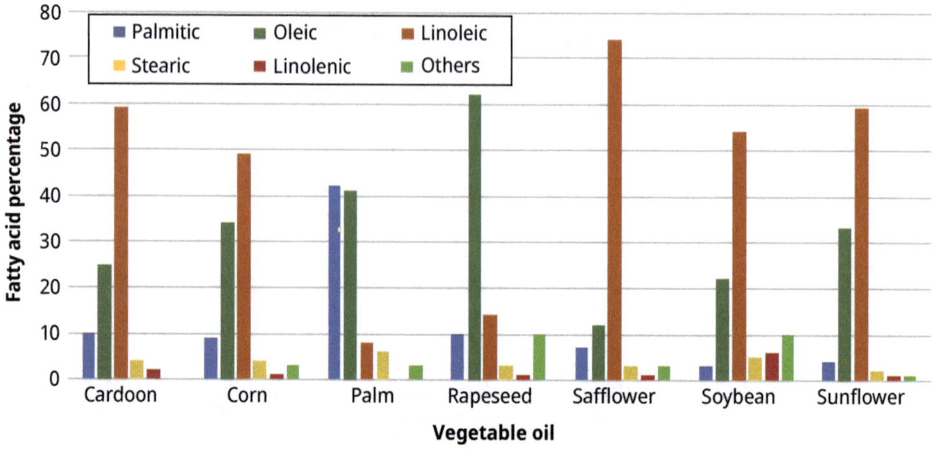

Fig. 2.8: Fatty acid profiles for typical vegetable oils.

Table 2.3 includes some of the main fatty acids in vegetable oils, and their main characteristics regarding oxidative stability. As it can be inferred from this table, a high percentage of linoleic or linolenic acid (with two and three unsaturations in their molecular structure, respectively) would promote a shorter or lower oxidative stability, whereas the presence of palmitic or oleic acid (with none and one unstaturations, respectively) would potentially increase the oxidative stability of the oil and, subsequently, the produced biodiesel or biolubricant. In that sense, there are plenty of studies that have proven this circumstance both in biodiesel and biolubricants, encouraging the use of raw materials with a higher ratio of oleic or palmitic acid [37, 41, 42]. Consequently, a suitable crop selection or even the use of genetically modified

Tab. 2.3: Comparison of different fatty acids and their oxidative stabilities.

Fatty acid	Lipid number (carbon:double bond)	Oxidative stability
Palmitic acid	16:0	üüü
Stearic acid	18:0	üüü
Oleic acid	18:1	üü
Linoleic acid	18:2	ü
Linolenic acid	18:3	ü

products (such as high-oleic vegetable oils) could be an interesting point to improve oxidative stability of biodiesel and biolubricants [43].

– **Mild reaction conditions**: Another aspect to be taken into consideration are the chemical conditions to produce biolubricants, mainly including catalyst addition and reaction temperature. As expected, the use of homogeneous catalysts as in the case of biodiesel production can be a cheap and suitable way to produce biolubricants. However, the purification process to remove surplus catalyst is more difficult and expensive in the second transesterification due to the high viscosity of the final product, with the subsequent presence of impurities. In that sense, the presence of catalysts during auto-oxidation processes plays a vital role, as they can catalyze this process, accelerating the above-mentioned problems related to viscosity and acidity increase. As a consequence, the use of mild reaction conditions could contribute to an improvement in oxidative stability, as long as a minimum conversion rate is obtained. Thus, the use of vacuum to remove methanol during the second transesterification could contribute to an increase in conversion, as the chemical balance promoted could go toward product generation. Consequently, both catalyst addition and reaction temperature could be reduced, contributing to an improvement in oxidative stability [44]. Also, the use of heterogeneous catalysts, already tested for biodiesel production to simplify the purification process of fatty acid methyl esters [45–49], could be an important way to avoid the use of homogeneous catalysts and their difficulty in separation processes from final biolubricant products. In that sense, the use of these products would decrease catalyst content in the final product, with the subsequent improvement in oxidative stability. Thus, a simple filtration would be enough to separate heterogeneous catalysts from biolubricant, but some challenges like the low reusability of heterogeneous catalysts need to be solved [46, 47, 50].

– **Antioxidant addition**: As commented in previous sections, the use of wastes (especially used frying oils) with difficult management to obtain biodiesel and biolubricants could be a very interesting point, but it should be noted that the quality of the raw material, which usually undergoes high temperature cycles during cooking (apart from other factors such as the presence of impurities), could be compromised, especially concerning oxidative stability. Nevertheless, this fact should not hinder the use of this oil as a raw material for the implementation of a biorefinery for biolubricant production, as the potential advantages outweigh this problem. Thus, the use of antioxidants, especially for this kind of raw material or other vegetable oils with low oxidative stability (due to the high presence of linoleic or linolenic acid, for instance), could be a viable alternative for the valorization of these raw materials. For this purpose, both natural antioxidants (included in extracts, for example) and synthetic antioxidants can be used. The former could contribute to the green chemistry policy, aiding in the reduction of synthetic compound addition in biodiesel orbiolubricant production. Thus, the use of these antioxidants is important, for instance, in biodiesel performance in diesel engines, having a positive impact [51]. However, their effectiveness or extraction could be difficult on many occasions, implying a very challenging and exciting research line in the literature, where these kinds of compounds have been proved in biodiesel samples obtained from different sources such as soybean [52–56]. On the other hand, synthetic antioxidants, which usually share many points in common with natural extracts regarding their molecular structure (Fig. 2.9), have been studied in the literature for the improvement of oxidative stability in biodiesel and biolubricants.

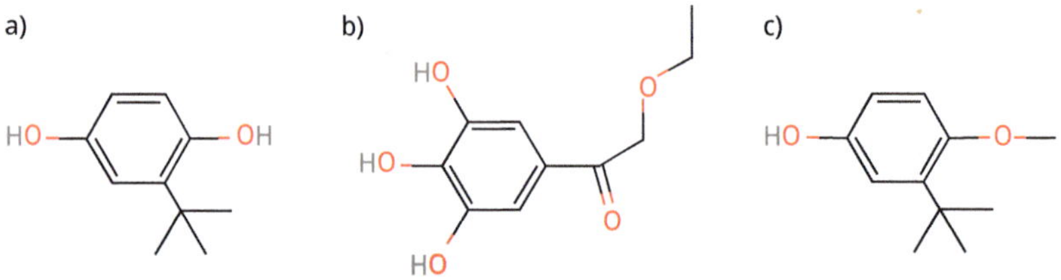

Fig. 2.9: Molecular structure of some antioxidants: (a) tert-butylhydroquinone (TBHQ); (b) propyl gallate (PG); and (c) butylated hydroxyanisole (BHA).

As observed in Fig. 2.9, three of the most studied antioxidants (not only for their addition in biolubricants, but also in other products like biodiesel [57–60], that is, tert-butylhydroquinone (TBHQ), propyl gallate (PG) and butylated hydroxyanisole (BHA)) are aromatic compounds, whose aromatic ring contributes to neutralize free radicals and, therefore, avoid the initiation and propagation stages affecting biodiesel and biolubricants. Consequently, the oxidative stability of the samples will considerably increase, implying a higher resistance to viscosity and acidity changes during storage or oxidation processes. Many studies have covered the use of some synthetic antioxidants, proving the high effectiveness and efficiency of these products [18, 61]. This point is very interesting in the case of the implementation of green policies, as the required concentration of these antioxidants is reduced, not exceeding 2,000 ppm in any case for a considerable improvement of the properties of biodiesel and biolubricants. In that sense, the use of such reduced amounts of antioxidants could offset the inconvenience of their use, as high-quality products with a high oxidative stability are obtained, even from wastes such as used frying oils.

The use of antioxidants in fatty acid esters presents an interesting scientific field, where many factors should be considered to optimize the effectiveness of these compounds in increasing the oxidative stability of products like biolubricants. For instance, there is a specific concentration range of antioxidants where their function seems to be optimal, showing a directly proportional increase in oxidative stability with antioxidant concentration. However, very low or excessive concentrations of antioxidants could not present a high efficiency, making necessary the study of a specific antioxidant and biolubricant to assess its effect on oxidative stability.

Another aspect of antioxidant addition is the possibility of synergistic effects in antioxidant mixtures, where not always the addition of two or more antioxidants assures a considerable increase in oxidative stability in biodiesel or biolubricant samples. Thus, previous studies have proven that the mixture of two antioxidants did not offer any improvement in their combined effectiveness, even showing an inhibition effect possibly due to the generation of less stable products during the neutralization of free radicals [17].

Finally, to understand the performance of antioxidants in biolubricants, some studies have been carried out to determine their concentration during storage or auto-oxidation processes. Thus, antioxidants like TBHQ or PG have been studied during the oxidation of biolubricants through analytical techniques like voltammetry, proving that the positive effect of antioxidants on biolubricants is, thanks to the direct neutralization of free radicals during oxidation, decreasing antioxidant concentration at the expense of keeping the quality of biolubricants [31].

2.5 Conclusions

In the light of the above, some conclusions can be drawn from this chapter, as follows:

- The use of vegetable oils for biolubricant production seems to be a promising alternative (or even a reality) for traditional lubricants, offering (on many occasions) better properties than the latter, such as higher flash or combustion points, which makes biolubricants a safe product during storage and shipping.
- However, some challenges should be addressed, as these products might present low oxidative stability (especially if a double transesterification process is carried out), altering their properties during storage.
- To solve the above-mentioned problem, the use of antioxidants, genetically modified raw materials or mild reaction conditions, could palliate the short oxidative stability of biolubricants.
- On the other hand, the use of energy crops for biolubricant production is a controversial issue, as these can be seen as a competence for food crops. In any case, some interesting crops like rapeseed or safflower could coexist or complement traditional crops, as they can take part in crop rotations. Moreover, these crops can be easily adapted to extreme climate conditions. In addition to all this, not all the biolubricant must be produced from raw materials coming from the countryside. Indeed, there is an increasing concern about frying oil management, which could be easily treated to obtain biolubricants, offering a smart and suitable solution.
- Forward-looking steps could imply the use of innovative catalysts to make biolubricant production more efficient compared to traditional practices, as well as the use of antioxidants or other measures to ensure high oxidative stability values that could keep the properties of biolubricants (mainly viscosity and acid number) during storage or auto-oxidation processes.

References

[1] Byrne J, Taminiau J. A review of sustainable energy utility and energy service utility concepts and applications: Realizing ecological and social sustainability with a community utility. Wiley Interdisciplinary Reviews: Energy and Environment. 2016;5:136–54.

[2] UN. Sustainable Development Goals, 2019.

[3] Vakulchuk R, Overland I, Scholten D. Renewable energy and geopolitics : A review. Renewable and Sustainable Energy Reviews. 2020;122:109547.

[4] Overland I, Bazilian M, IlimbekUulu T, Vakulchuk R, Westphal K. The GeGaLo index: Geopolitical gains and losses after energy transition. Energy Strategy Reviews. 2019;26:100406.

[5] Gururaja J. Energy for sustainable development: Review of national and international energy policies. Natural Resources Forum. 2003;27:53–67.

[6] Ajala OE, Aberuagba F, Odetoye TE, Ajala AM. Biodiesel: Sustainable energy replacement to petroleum-based diesel Fuel – A review. ChemBioEng Reviews. 2015;2:145–56.

[7] Mobarak HM, Niza Mohamad E, Masjuki HH, Kalam MA, Al Mahmud KAH, Habibullah M, Ashraful AM. The prospects of biolubricants as alternatives in automotive applications. Renewable and Sustainable Energy Reviews. 2014;33:34–43.

[8] Ho CK, McAuley KB, Peppley BA. Biolubricants through renewable hydrocarbons: A perspective for new opportunities. Renewable and Sustainable Energy Reviews. 2019;113:109261.

[9] Mobarak HM, Niza Mohamad E, Masjuki HH, Kalam MA, Al Mahmud KAH, Habibullah M, Ashraful AM. The prospects of biolubricants as alternatives in automotive applications. Renewable and Sustainable Energy Reviews. 2014;33:34–43.

[10] Chan CH, Tang SW, Mohd NK, et al. Tribological behavior of biolubricant base stocks and additives. Renewable and Sustainable Energy Reviews. 2018;93:145–57.

[11] Kania D, Yunus R, Omar R, Abdul Rashid S, Mohamad Jan B. A review of biolubricants in drilling fluids: Recent research, performance, and applications. Journal of Petroleum Science and Engineering. 2015;135:177–84.

[12] Scopus. Scopus. https://www.scopus.com/home.uri.

[13] Cecilia JA, Plata DB, Saboya RMA, Fmt DL, Cavalcante CL, Rodríguez-Castellón E. An overview of the biolubricant production process: Challenges and future perspectives. Processes. 2020;8:1–24.

[14] Ghosh N, Rhithuparna D, Rokhum SL, Halder G. Ethical issues pertaining to sustainable biodiesel synthesis over trans/esterification process. Sustainable Chemistry and Pharmacy. 2023. https://doi.org/10.1016/j.scp.2023.101123

[15] Devarajan Y, Munuswamy DB, Subbiah G, Vellaiyan S, Nagappan B, Varuvel EG, Thangaraja J. Inedible oil feedstocks for biodiesel production: A review of production technologies and physicochemical properties. Sustainable Chemistry and Pharmacy. 2022. https://doi.org/10.1016/j.scp.2022.100840

[16] Nogales-Delgado S, Sánchez N, Encinar JM. Valorization of cynara Cardunculus L. Oil as the basis of a biorefinery for biodiesel and biolubricant production. Energies (Basel). 2020. https://doi.org/10.3390/en13195085

[17] Nogales-Delgado S, Encinar JM, González JF. Safflower biodiesel: Improvement of its oxidative stability by using BHA and TBHQ. Energies (Basel). 2019. https://doi.org/10.3390/en12101940

[18] Encinar JM, Nogales-Delgado S, Álvez-Medina CM. High oleic safflower biolubricant through double transesterification with methanol and pentaerythritol: Production, characterization, and antioxidant addition. Arabian Journal of Chemistry. 2022;15:103796.

[19] Encinar JM, Nogales-Delgado S, Pinilla A. Biolubricant production through double transesterification: Reactor design for the implementation of a biorefinery based on rapeseed. Processes. 2021;9:1224.

[20] Encinar JM, Nogales-Delgado S, Sánchez N, González JF. Biolubricants from rapeseed and castor oil transesterification by using titanium isopropoxide as a catalyst: Production and characterization. Catalysts. 2020. https://doi.org/10.3390/catal10040366
[21] Fonseca JM, Teleken JG, De Cinque Almeida V, Da Silva C. Biodiesel from waste frying oils: Methods of production and purification. Energy Conversion and Management. 2019;184:205–18.
[22] Ouanji F, Nachid M, Kacimi M, Liotta LF, Puleo F, Ziyad M. Small scale biodiesel synthesis from waste frying oil and crude methanol in Morocco. Chinese Journal of Chemical Engineering. 2016;24:1178–85.
[23] Berrios M, Martín MA, Chica AF, Martín A. Study of esterification and transesterification in biodiesel production from used frying oils in a closed system. Chemical Engineering Journal. 2010;160:473–79.
[24] Roy T, Agarwal AK, Chandra Y. A cleaner route of biodiesel production from waste frying oil using novel potassium tin oxide catalyst. A Smart Liquid-waste Management. 2021;135:243–55.
[25] Harabi M, Bouguerra SN, Marrakchi F, Chrysikou LP, Bezergianni S, Bouaziz M. Biodiesel and crude glycerol from waste frying oil: Production, characterization and evaluation of biodiesel oxidative stability with diesel blends. Sustainability (Switzerland). 2019. https://doi.org/10.3390/su11071937
[26] Encinar JM, Nogales S, González JF. Biodiesel and biolubricant production from different vegetable oils through transesterification. Engineering Reports. 2020;2(12):1–10.
[27] Ameen NHA, Durak E. Study of the tribological properties the mixture of soybean oil and used (waste) frying oil fatty acid methyl ester under boundary lubrication conditions. Renew Energy. 2020;145:1730–47.
[28] Statista. U.S. population: Most abundant purposes for the usage of salad or cooking oil from 2011 to 2020. Based on MRI-Simmons data; among users of salad or cooking oil, 2023.
[29] McNutt J, He QS. Development of biolubricants from vegetable oils via chemical modification. Journal of Industrial and Engineering Chemistry. 2016;36:1–12.
[30] Owuna FJ, Dabai MU, Sokoto MA, Dangoggo SM, Bagudo BU, Birnin-Yauri UA, Hassan LG, Sada I, Abubakar AL, Jibrin MS. Chemical modification of vegetable oils for the production of biolubricants using trimethylolpropane: A review. Egyptian Journal of Petroleum. 2020;29:75–82.
[31] Nogales-Delgado S, Cabanillas AG, ÁG Romero, Encinar Martín JM. Monitoring tert-butylhydroquinone content and its effect on a biolubricant during oxidation. Molecules. 2022;27:8931.
[32] Sia CB, Kansedo J, Tan YH, Lee KT. Evaluation on biodiesel cold flow properties, oxidative stability and enhancement strategies: A review. Biocatalysis and Agricultural Biotechnology. 2020;24:101514.
[33] Liberato V, Benevenuti C, Coelho F, Botelho A, Amaral P, Pereira N, Ferreira T. Clostridium sp. as bio-catalyst for fuels and chemicals production in a biorefinery context. Catalysts. 2019;9(11):1–37.
[34] Palmeros Parada M, Osseweijer P, Posada Duque JA. Sustainable biorefineries, an analysis of practices for incorporating sustainability in biorefinery design. Industrial Crops and Products. 2017;106:105–23.
[35] Ebrahimian E, Denayer JFM, Aghbashlo M, Tabatabaei M, Karimi K. Biomethane and biodiesel production from sunflower crop: A biorefinery perspective. Renew Energy. 2022;200:1352–61.
[36] Granjo JFO, Duarte BPM, Oliveira NMC. Integrated production of biodiesel in a soybean biorefinery: Modeling, simulation and economical assessment. Energy. 2017. https://doi.org/10.1016/j.energy.2017.03.167
[37] Encinar JM, Nogales S, González JF. 2019 Biorefinery based on different vegetable oils: Characterization of biodiesel and biolubricants. 3rd International Conference in Engineering Applications (ICEA). 2019. https://doi.org/10.1109/CEAP.2019.8883468
[38] Focke WW, Van Der Westhuizen I, Oosthuysen X. Biodiesel oxidative stability from Rancimat data. Thermochimica Acta. 2016. https://doi.org/10.1016/j.tca.2016.03.023
[39] Kumar N. Oxidative stability of biodiesel: Causes, effects and prevention. Fuel. 2017. https://doi.org/10.1016/j.fuel.2016.11.001

[40] Tang H, De Guzman RC, Salley SO, Ng SKY. The oxidative stability of biodiesel: Effects of FAME composition and antioxidant. Lipid Technology. 2008;20:249–52.

[41] Martínez G, Sánchez N, Encinar JM, González JF. Fuel properties of biodiesel from vegetable oils and oil mixtures. Influence of methyl esters distribution. Biomass Bioenergy. 2014. https://doi.org/10.1016/j.biombioe.2014.01.034

[42] Salimon J, Salih N, Yousif E. Biolubricants: Raw materials, chemical modifications and environmental benefits. European Journal of Lipid Science and Technology. 2010;112:519–30.

[43] Nogales-Delgado S, Encinar JM, Cortés Á G. High oleic safflower oil as a feedstock for stable biodiesel and biolubricant production. Industrial Crops and Products. 2021;170:113701.

[44] Nogales-Delgado S, Encinar Martín JM, Sánchez Ocaña M. Use of mild reaction conditions to improve quality parameters and sustainability during biolubricant production. Biomass Bioenergy. 2022;161:106456.

[45] Encinar JM, González JF, Pardal A, Martínez G. Rape oil transesterification over heterogeneous catalysts. Fuel Processing Technology. 2010. https://doi.org/10.1016/j.fuproc.2010.05.034

[46] Di Serio M, Tesser R, Pengmei L, Santacesaria E. Heterogeneous catalysts for biodiesel production. Energy & Fuels. 2008;22:207–17.

[47] Jayakumar M, Karmegam N, Gundupalli MP, BizunehGebeyehu K, Tessema Asfaw B, Chang SW, Ravindran B, Kumar Awasthi M. Heterogeneous base catalysts: Synthesis and application for biodiesel production – A review. Bioresource Technology. 2021;331:125054.

[48] Petrov L, Kumbilieva K. Kinetic studies of catalyst deactivation in heterogeneous catalysis. Chemical Informatics. 2006. https://doi.org/10.1002/chin.200627269

[49] Kawashima A, Matsubara K, Honda K. Development of heterogeneous base catalysts for biodiesel production. Bioresource Technology. 2008;99:3439–43.

[50] Masudi A, Muraza O. Vegetable oil to biolubricants: Review on advanced porous catalysts. Energy and Fuels. 2018;32:10295–310.

[51] Hosseinzadeh-Bandbafha H, Kumar D, Singh B, Shahbeig H, Lam SS, Aghbashlo M, Tabatabaei M. Biodiesel antioxidants and their impact on the behavior of diesel engines: A comprehensive review. Fuel Processing Technology. 2022;232:107264.

[52] De Sousa LS, De Moura CVR, De Oliveira JE, De Moura EM. Use of natural antioxidants in soybean biodiesel. Fuel. 2014;134:420–28.

[53] Silva de Sousa L, Verônicarodarte de Moura C, de Moura E M. Action of natural antioxidants on the oxidative stability of soy biodiesel during storage. Fuel. 2020. https://doi.org/10.1016/j.fuel.2020.119632

[54] Tang H, Wang A, Salley SO, Ng KYS. The effect of natural and synthetic antioxidants on the oxidative stability of biodiesel. Journal of the American Oil Chemists Society. 2008;85(4):1–10.

[55] Jemima Romola CV, Meganaharshini M, Rigby SP, Ganesh Moorthy I, Shyam Kumar R, Karthikumar S. A comprehensive review of the selection of natural and synthetic antioxidants to enhance the oxidative stability of biodiesel. Renewable and Sustainable Energy Reviews. 2021;145:111109.

[56] Nogales-Delgado S, Encinar JM, Guiberteau A, Márquez S. The effect of antioxidants on corn and sunflower biodiesel properties under extreme oxidation conditions. JAOCS, Journal of the American Oil Chemists' Society. 2019. https://doi.org/10.1002/aocs.12288

[57] Zhou J, Xiong Y, Shi Y. Antioxidant consumption kinetics and shelf-life prediction for biodiesel stabilized with antioxidants using the Rancimat method. Energy and Fuels. 2016;30:10534–42.

[58] Ramalingam S, Rajendran S, Ganesan P, Govindasamy M. Effect of operating parameters and antioxidant additives with biodiesels to improve the performance and reducing the emissions in a compression ignition engine – A review. Renewable and Sustainable Energy Reviews. 2018. https://doi.org/10.1016/j.rser.2017.08.026

[59] Van der Westhuizen I, Focke WW. Stabilizing sunflower biodiesel with synthetic antioxidant blends. Fuel. 2018;219:126–31.

[60] Souza AG, Medeiros ML, Cordeiro AMMT, Queiroz N, Soledade LEB, Souza AL. Efficient antioxidant formulations for use in biodiesel. Energy and Fuels. 2014;28:1074–80.

[61] Quinchia LA, Delgado MA, Valencia C, Franco JM, Gallegos C. Natural and synthetic antioxidant additives for improving the performance of new biolubricant formulations. Journal of Agricultural and Food Chemistry. 2011;59:12917–24.

Lokesh Kumar, Anupama, Dheeraj Ahuja

3 A short review of renewable castor oil-based polyurethane (PU)

3.1 Background of polyurethane (PU)

Polyurethanes (PUs) are polymers containing urethane linkage and are being utilized in a variety of applications such as coatings [1–3], foams [4], adhesives [5], sealants [6], and biomedical uses [7, 8] because of their durability, elasticity, fatigue resistance, compliance, biostability, biodegradability, and advantage of tunable structure from a variety of available precursors. They are the most adaptable class of polymeric materials. They introduced the polyaddition reaction of isocyanate group and diols to produce urethane (Fig. 3.1). Any polymers that contain urethane, urea, or other isocyanate-derived groups are included in the group of PUs. PUs are characterized by carbamate groups ($-NHCO_2$) in their molecular backbone [9].

O
||
>NH—C—O—

Fig. 3.1: Structure of urethane linkage.

PU has a lot of distinctive advantages and is a substance that is frequently used. Various properties of a polymer depend on its molecular weight, crystallinity, tensile strength, cross-linking, and so on. These properties can be tuned in a wide range by using different processing routes and techniques. PUs exist in a variety of forms like flexible and rigid foams, fibers [10], paints and coatings [11–13], adhesives [14, 15], and elastomers; thus these have found a number of useful industrial applications in diverse areas such as construction, furniture, thermal insulation, footwear, automotive, apparel, marine, packaging, and electronics.

3.2 Chemistry of PU

PU is generally prepared by the reaction of a long-chain diol with an excess of diisocyanate:

$$R_1(NCO)_2 + HO - {}^{\sim\sim}R_2{}^{\sim\sim} - OH \rightarrow OCN - R_1 - NHCOO^{\sim\sim\sim}OCONH - R_2 - NCO \quad (3.1)$$

where R_1 is an aromatic, aliphatic, or cycloaliphatic radical, and R_2 is usually an aliphatic group, derived from polyether or polyester [16]. The reaction between diisocyanates and diols, proceeding in the presence of catalysts [17], is a basic reaction for the synthesis of various PUs. The "prepolymer" produced in the reaction can react sepa-

https://doi.org/10.1515/9783111145136-003

rately with diols and diamines of low molecular weight to cause further extension or polymerization (curing of the prepolymer). On reaction with water, the prepolymer can also give foams. The properties of the PU resin are controlled by the choice and ratio of the diisocyanate and polyol.

3.3 Raw materials

Polyols and diisocyanates are the two most important monomers required for the preparation of PU. Polyols are high-molecular-weight compounds that act as a soft segment, while polydiisocyantes act as a hard segment. Apart from these two main components, chain extenders, that is, low-molecular-weight diols and diamines, are also used for the extension of reaction for proper curing of the PU. Apart from these catalysts, different blowing agents, solvents, and flame-retardant capping agents are also added for the production of different types of PUs such as foams, elastomers, coatings, paints, shoe soles, and so on.

3.3.1 Polyols

High-molecular-weight compounds containing two or more reactive hydroxyl groups synthesized from an initiator and monomeric building blocks are known as polyols. A wide range of polyols with different chemical structures and hydroxyl numbers are available for PU synthesis. They can be classified as polyester polyols, polyether polyols, dihydroxy and polyhydroxy alcohols, oil-based polyols, biobased polyols, lactide-based polyols, lignin-based polyols, starch-based polyols, and other biobased polyols. The basic chemical structure and the average number of hydroxyl groups available in polyol provide extreme flexibility to maneuver the formulations to obtain the desired characteristics of the final product. Polyols are generally classified into two main categories, that is, polyester polyols and polyether polyols.

3.3.1.1 Polyether polyols

Polyols based on polyether are the most important commercial compounds used to synthesize PU. In 1962, Saunders et al. designed poly(oxytetramethylene) glycol derived from tetrahydrofuran as the first polyether for preparing PU.

For the preparation of polyether polyols, alkylene oxides are subjected to addition reaction with alcohols or amines using anionic or cationic mechanism [18]. For example, for synthesis of polyether polyols, propylene oxide and ethylene oxide are the two most commonly used alkylene oxides. Polyether polyols derived from alkylene oxides are less reactive because of the presence of hydroxyl group at the secondary position [18]. Therefore, nitrogen is added for fast curing during the preparation of polyol [19, 20]. They can also be obtained by polycondensation of propane-1,3-diol.

3.3.1.2 Polyester polyols

Polyester polyols are synthesized by polycondensation reaction of acidic and hydroxy monomers in their repeated form known as polyacid and polyhydroxy. Sometimes polyacid is replaced by polyanhydride because of their incompatibility with condensation and relatively high cost. Triols are used for introducing the branching that enhances not only the hardness and rigidity but also the heat and chemical resistance of the PUs. Utilization of polyester polyols for the development of PU results in a high level of tensile and flexural strength along with good abrasion and oil resistance. But the major drawback with them is that they can be easily hydrolyzed.

3.3.2 Diisocyanates

An organic compound with a functional group of one atom each of nitrogen, carbon, and oxygen arranged as –N = C = O is referred to as "isocyanate," and it can have more than one isocyanate group. Isocyanates with more than two isocyanate groups are termed "diisocyanates" and are important raw materials for the synthesis of PUs. The reaction of a diisocyanate with a compound containing two or more hydroxyl groups results in the formation of PU, while with a compound containing two or more amine groups produces polyureas. In addition, the reaction of isocyanate with water results in the formation of carbon dioxide that is used as a blowing agent to produce PU foams. NCO equivalent weight, %NCO, and functionality of isocyanates are the key characteristics that affect the properties of PU systems. Among all the three parameters, isocyanate index is an important factor that influences the PU properties. Isocyanate index is the amount of isocyanate consumed relative to the amount that is theoretically required. For attaining a higher hardness, the isocyanate index shall be higher.

Diisocyanates can be classified as aromatic and aliphatic isocyanates. Both aromatic as well as aliphatic isocyanates can be used to synthesize PUs. The two most widely utilized isocyanates for the commercial development of PUs are 2,4- and 2,6-toluene diisocyanate (TDI) and 4,4′-diphenylmethane diisocyanate (MDI). The structure of both the isocyanates is shown in Fig. 3.2(a) and (b).

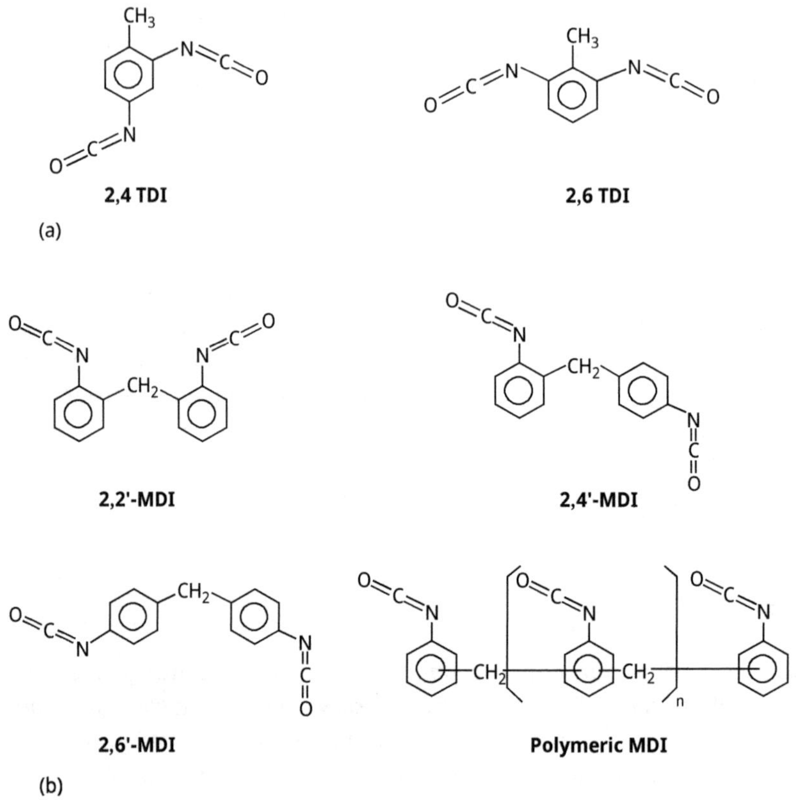

Fig. 3.2: Structure of different types of (a) TDI and (b) MDI used for the preparation of PU.

3.3.3 Chain extenders

Incorporating chain extenders, such as diols of low molecular weight, enhances the elastomeric properties of the resulting PUs, because the small diols react with diisocyanates and form hard domains to serve as the physical cross-link for the PU systems. Some of the most commonly used chain extenders in biodegradable PU formulations are butanediol (BDO), ethylene glycol (EG), 1,3-butanediol, 2,2-dimethyl-1,3-propanediol, trimethylopropane, glycerol, 1,2,6-hexanetriol, 1,4-cyclohexanedimethanol (CHDM), 1,2-ethanediamine (ED), 1,4-butanediamine (BDA), 1,6-hexanediol, 1,5-heptanediol, and 2-amino-1-butanol (ABDO). Aromatic chain extenders produce stiffer PUs in comparison to aliphatic chain extenders. Also, carbon atoms present in the chain extender significantly affect the structure of the hard segment with a degree of phase separation and melting temperatures.

3.3.4 Foaming agents

Another important component that is mainly used for the manufacturing of PU foams is the foaming agent, which creates gas bubbles inside the structure. Water is a popular nontoxic blowing agent due to its low cost and environmental friendliness. Additionally, CO_2 that is formed from water helps to enhance various properties of PU foam such as density, strength, and porosity. Blowing agents, other than water, include chlorofluorocarbons, which are in liquid phase at standard state but turn into vapor phase after absorbing a sufficient amount of heat from the process. Since chlorofluorocarbons are considered a major cause of destruction of the ozone layer, their applications and production have been rigorously constrained.

3.3.5 Catalysts

Catalysts used for the development of PU are generally classified into two types: amine compounds and organometallic complexes. Tertiary amines are preferred over organometallic complexes for the manufacturing of based catalyst are used as they furnish the change in the degree of phase mixing, hard domain structure and hydrogen bonding.

3.4 Transformation toward renewable castor oil (CO)-based PUs

The major limitation with the PU produced from fossil fuels is its nonbiodegradable nature, which results in disturbing and damaging the ecosystem of nature [18]. This major limitation can be overcome by using biodegradable PUs. Encouraging regulatory policies toward manufacturing biobased products coupled with rigorous environmental regulations for manufacturing petrochemicals and synthetic polymers has positively impacted biobased PU market. Biodegradability of PUs is generally achieved by incorporating labile and hydrolyzable moieties into the polymer backbone [19–22]. The most common method for achieving this goal is the application of polyols (soft segments) with hydrolyzable bonds as the starting materials for the preparation of PUs.

Polyester and polyether polyols are the most widely used polyols for the synthesis of PUs; however, both the polyols are petrobased that result in nonbiodegradable PU. The first approach toward the development of biobased PU is to replace polyols with biobased polyols. Biobased polyether polyols can be obtained using sorbitol but those are also partially renewable, while the biobased polyester polyols are derived from the condensation of biobased dicarboxylic acids, such as adipic acid or succinic acid, with biobased polyols (propane-1, 3-diol).

However, both of these obtained biobased polyols are partially renewable and account for less content of renewable resources in the final product PU. Hence, a need for the development of polyols that are totally biobased was demanded.

Furthermore, it is reported that the biodegradability in PU is generally achieved by incorporating labile and hydrolyzable moieties into the polymer backbone [19–22]. The most common method for fulfilling this goal is the application of polyols (soft segments) with hydrolyzable bonds as starting materials for the preparation of PUs.

Several hydroxyl-terminated polymers, such as polycaprolactone, polyalkylene adipate, polylactides, and polyglycolides, have been used for the synthesis of hydrolytically degradable PUs [23–29] for various applications because of their biodegradability and these being independent of petroleum and nontoxic degradation products. In addition, using nonaromatic diisocyanate can also contribute to biocompatibility properties of PUs.

Some of the polymeric materials based on bio-renewable resources include polysaccharides, such as starch, chitin/chitosan, cellulose, alginate, carbohydrate polymers and animal protein-based biopolymers such as gelatin, wool, silk, and collagen and oil-based PUs [30].

Starch, after cellulose, is one of the most important water-insoluble granules obtained from plants, such as corn, wheat grains, cassava, cereal, rice, and potato, other than stems, roots, bulbs or legumes, and nuts. Amylose and amylopectin linked together with α-d-[1–4] and/or α-d [1–6] linkages are two main polysaccharide units of starch. Biodegradability and structural amenability are two important characteristics that make starch versatile and unique material for use in various industries ranging like washing detergents. Also, easy availability, less cost, and ecocompatability with nature make it a potential raw material to be used for the synthesis of biodegradable polymers. Starch can be used directly as a polyol for the synthesis of PU. It can also be used as a filler for the synthesis of PU composites with enhanced mechanical properties. Starch can directly be used as a polyol as it contains amylose and amylopectin macromolecules that react with hot water to form inter and intermolecular crosslinks that lead to the formation of a molecular network, resulting in the formation of a film after the water is evaporated [31, 32].

Lignin is being explored as a raw material for the synthesis of PU because of its random noncrystalline network structure, leading to excellent mechanical properties and thermal stability. Moreover, aliphatic and phenolic hydroxyl functionalities of lignin provide good reacting sites for isocyanates [33] and the inherent properties of lignin contribute toward the improvement of moisture and flame resistance of PUs [34]. PU can be synthesized utilizing lignin either by direct incorporation of lignin into PU matrix without any chemical modification or in combination with other polyols or by chemically modifying lignin by esterification/etherification to make hydroxyl group more readily available for the reaction.

Other biobased polyols include glycerol, xylitol, sorbitol, cellulose, and so on. Glycerol is an important starter polyether polyol for the preparation of flexible foams. D-glucose obtained by acid hydrolysis of starch is another important compound used for the synthesis of sorbitol (another significant polyether polyols). Cellulose can directly be used as a polyol for PU formation. Kumari et al. [35] directly utilized cellulose as a polyol along with lignin for synthesizing PU foam for wastewater treatment. Gupta et al. [36] synthesized a limenene-based polyol using thiol-ene click chemistry for the development of rigid PU foams.

3.5 Oil-based PUs

The preparation of polymers from renewable sources, such as vegetable oil-based materials, is currently receiving increasing attention because of the economic and environmental concerns [37, 38]. Furthermore, PUs prepared from vegetable oils possess excellent mechanical and thermal properties and are more ecofriendly when compared to petroleum-based products [39–41].

Vegetable oils are triglycerides with at least one unsaturated fatty acid in their chemical structure [42, 43] and contain various reactive sites such as double bonds and ester groups that make them important resources for polyol. To use vegetables as raw materials for the development of PU, it is necessary to functionalize them to form polyols. Some of the most common techniques utilized for the functionalization of vegetable oil to polyols include epoxidation and ring-opening reaction with halo acids or alcohols, ozonolysis, and hydration. Rapeseed, tung, linseed, canola, sunflower, soybean, and CO are the most commonly used vegetable oils that have either directly or by chemical modification been used as a polyol precursor for PU synthesis.

3.6 Castor oil-based PUs

Among various vegetable oils, CO, shown in Fig. 3.3, is a promising raw material due to its low cost, low toxicity, and availability as a renewable agricultural resource. It is a viscous, pale yellow nonvolatile and nondrying oil with a bland taste and is sometimes used as a purgative. It has a slight characteristic odor while the crude oil tastes slightly acrid with a nauseating after-taste. When compared to other vegetable oils, it has a good shelf-life, and it does not turn rancid unless subjected to excessive heat. India is the world's largest exporter of CO; other major producers are China and Brazil.

Its major constituent, ricinoleic acid (12-hydroxy-*cis*-9-octadecenoic acid), is a hydroxyl-containing fatty acid [44]. CO can be used directly as a raw material for the preparation of PUs without any future modification [45–48].

CO is chosen for the fabrication of elastomeric PU because of many special structural attributes that lend themselves as a raw material to a plethora of organic reactions. CO has an overall functionality of about 2.7 and is therefore able to act as a

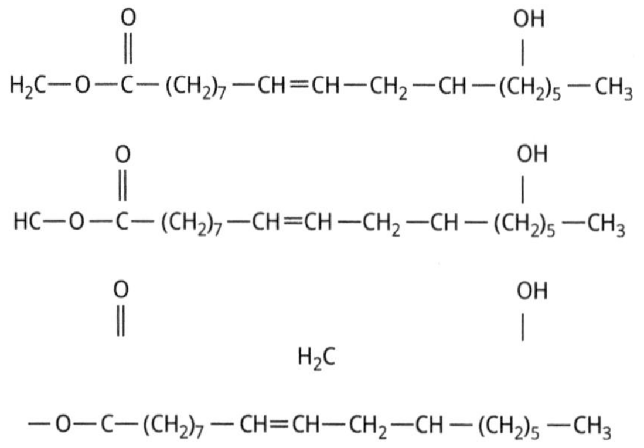

Fig. 3.3: Castor oil chemical framework [44].

proton donor because of its hydroxyl groups, as well as a cross-linking agent because of its functionality. The unique dynamics between the soft and hard segments of PUs segregate into microdomains that result in their excellent elastomeric properties.

The PUs obtained from CO are typically water resistant and flexible due to the presence of a long fatty acid chain and lend themselves to thermosetting-type structures due to their trifunctional nature. The long pendant chains of the fatty acids impart flexibility and hydrolysis resistance to the network and the double bond present in the CO serves as a grafting center. Urethane polymers exhibit excellent abrasion resistance. CO has been used for the synthesis of PU foams, artificial bones, and semi-interpenetrating polymer networks [49–52].

However, disadvantages of the direct use of CO as a polyol include low hydroxyl number, leading to inherently low modulus materials, a sluggish rate of curing of the secondary hydroxyl groups 2, and structural irregularity due to steric hindrance caused by the long pendant fatty acid chains during urethane formation, resulting in low tear strength [53]. These disadvantages can be overcome by transesterification of CO with other polyols with high functionality and high hydroxyl number or transamidation with polyamines or alkanolamines leading to the formation of rigid PU with good physical–mechanical properties [53].

3.7 Applications of polyurethanes

PUs possess potential applications in various diversified industries such as coatings, drug and gene delivery, paints, automotive, additives, sealants, wastewater treatment, sorption, macromolecular building, supramolecular science, and so on because their physical and chemical properties can be varied in a controlled way by changing the percentage of the composition of soft and hard segments. PU coatings are widely used due to their excellent resistance against corrosion, chemical and abrasion along with toughness, and low temperature flexibility, and so these can be used in various applications in diverse areas ranging from automobile to industrial maintenance [54]. PUs are also used for the synthesis of dashboards and other plastic parts because of their lightweight and higher mechanical strength that reduce the overall weight of car, leading to higher fuel efficiency and thus providing enhanced environmental protection [55].

In addition, PUs are used in many other areas such as in footwear, furniture, bedding, refrigeration, and freezers.

3.8 Conclusion

PUs, because of their ability to tune properties according to the requirement by variation in the molar ratio of soft and hard segments, can be utilized in almost all areas such as automobiles, paints, coatings, furniture, footwear, biomedical, construction, and so on. PUs derived from CO provide biodegradability in addition to enhancement in their physical and chemical properties, making them suitable for manifold applications ranging from various household products to aviation products. This chapter discussed the current progress in the area of CO-based PUs along with their chemistry and applications. A critical appraisal of the existing literature affirms various opportunities in the area of CO-based PUs. Presently, CO-based polyurethanes have some disadvantages like sluggish rate of curing. However, probability of using nanomaterials in the arena of CO-based PUs has made their future secure.

References

[1] Noreen A, Zia KM, Zuber M, Tabasum S, Saif MJ. Recent trends in environmentally friendly waterborne polyurethane coatings: A review. Korean Journal of Chemical Engineering. 2016;33:388–400.

[2] Shen R, Long M, Lei C, Dong L, Yu G, Tang J. Anticorrosive waterborne polyurethane coatings derived from castor oil and renewable diols. Chemical Engineering Journal. 2022;433:134470.

[3] Rabbani S, Bakhshandeh E, Jafari R, Momen G. Superhydrophobic and icephobic polyurethane coatings: Fundamentals, progress, challenges and opportunities. Progress in Organic Coatings. 2022;165:106715.

[4] Kaikade DS, Sabnis AS. Polyurethane foams from vegetable oil-based polyols: A review. Polymer Bulletin. 2023;80(3):2239–61.

[5] Tenorio-Alfonso A, Sánchez MC, Franco JM. Impact of the processing method on the properties of castor oil/cellulose acetate polyurethane adhesives for bonding wood. International Journal of Adhesion and Adhesives. 2022;116:103153.

[6] Shi S, Ma T, Gu L, Zhang Y. Thermal behaviors, interfacial microstructure and molecular orientation of shape memory polyurethane/SiO2 based sealant for concrete pavement. Polymers (Basel). 2022;14(16):3336.

[7] Dai S, Yue S, Ning Z, Jiang N, Gan Z. Polydopamine nanoparticle-reinforced near-infrared light-triggered shape memory polycaprolactone–polydopamine polyurethane for biomedical implant applications. ACS Applied Materials and Interfaces. 2022;14(12):14668–76.
[8] Tan RYH, Lee CS, Pichika MR, Cheng SF, Lam KY. PH responsive polyurethane for the advancement of biomedical and drug delivery. Polymers (Basel). 2022;14(9):1672.
[9] Staley HB, David DJ. Analytical Chemistry of the Polyurethanes. New York: Robert E. Krieger, 1979.
[10] El-Molla MM. Synthesis of polyurethane acrylate oligomers as aqueous UV-curable binder for inks of ink jet in textile printing and pigment dyeing. Dye Pigment. 2007;74(2):371–79.
[11] Kong X, Liu G, Qi H, Curtis JM. Preparation and characterization of high-solid polyurethane coating systems based on vegetable oil derived polyols. Progress in Organic Coatings. Internet 2013;76(9):1151–60. Available from: http://dx.doi.org/10.1016/j.porgcoat.2013.03.019
[12] Patel CJ, Mannari V. Air-drying bio-based polyurethane dispersion from cardanol: Synthesis and characterization of coatings. Progress in Organic Coatings. 2014;77(5):997–1006.
[13] Rajput SD, Hundiwale DG, Mahulikar PP, Gite VV. Fatty acids based transparent polyurethane films and coatings. Progress in Organic Coatings. 2014;77(9):1360–68.
[14] Ates B, Koytepe S, Karaaslan MG, Balcioglu S, Gulgen S. Biodegradable non-aromatic adhesive polyurethanes based on disaccharides for medical applications. International Journal of Adhesion and Adhesives. Internet 2014;49:90–96. Available from: http://www.sciencedirect.com/science/article/pii/S0143749613002236
[15] Ren D, Frazier CE. Wood/adhesive interactions and the phase morphology of moisture-cure polyurethane wood adhesives. International Journal of Adhesion and Adhesives. 2012;34:55–61.
[16] Król P. Generalization of kinetics in the reaction of isocyanates and polyols for modeling a process-yielding linear polyurethane, 1. Journal of Applied Polymer Science. 1995;57(6):739–49.
[17] Woods G. The ICI Polyurethanes Book. ICI Polyurethanes. Wiley, 1987.
[18] Xiao H, Koleske J. Polyurethane coatings. Paint Coat Test Man 15th Ed Gardner-Sward Handb. 2012;102–12.
[19] Okada M. Chemical syntheses of biodegradable polymers. Progress in Polymer Science. 2002;27(1):87–133.
[20] Howard GT. Biodegradation of polyurethane: A review. International Biodeterioration and Biodegradation. 2002;49(4):245–52.
[21] Chandra R, Rustgi R. Biodegradable polymers. Progress in Polymer Science. 1998;23(7):1273–335.
[22] Pavlova M, Draganova M. Biocompatible and biodegradable polyurethane polymers. Biomaterials. 1993;14(13):1024–29.
[23] Yeganeh H, Jamshidi H, Jamshidi S. Synthesis and properties of novel biodegradable poly (ε-caprolactone)/poly (ethylene glycol)-based polyurethane elastomers. Polymer International. 2007;56(1):41–49.
[24] Gorna K, Gogolewski S. Biodegradable polyurethanes for implants. II. In vitro degradation and calcification of materials from poly (ϵ-caprolactone)–poly (ethylene oxide) diols and various chain extenders. Journal of Biomedical Materials Research, Part A. 2002;60(4):592–606.
[25] Guan J, Sacks MS, Beckman EJ, Wagner WR. Biodegradable poly (ether ester urethane) urea elastomers based on poly (ether ester) triblock copolymers and putrescine: Synthesis, characterization and cytocompatibility. Biomaterials. 2004;25(1):85–96.
[26] Kylmä J, Seppälä JV. Synthesis and characterization of a biodegradable thermoplastic poly (ester– urethane) elastomer. Macromolecules. 1997;30(10):2876–82.
[27] Lee S-I, S-c Y, Lee Y-S. Degradable polyurethanes containing poly (butylene succinate) and poly (ethylene glycol). Polymer Degradation and Stability. 2001;72(1):81–87.
[28] Guan J, Sacks MS, Beckman EJ, Wagner WR. Synthesis, characterization, and cytocompatibility of elastomeric, biodegradable poly (ester-urethane) ureas based on poly (caprolactone) and putrescine. Journal of Biomedical Materials Research, Part A. 2002;61(3):493–503.

[29] Tang YW, Labow RS, Revenko I, Santerre JP. Influence of surface morphology and chemistry on the enzyme catalyzed biodegradation of polycarbonate-urethanes. Journal of Biomaterials Science, Polymer Edition. 2002;13(4):463–83.
[30] Xiao HX, Koleske JV. Polyurethane Coatings. Paint and Coating Testing Manual. 2012;102–12.
[31] Miles MJ, Morris VJ, Orford PD, Ring SG. The roles of amylose and amylopectin in the gelation and retrogradation of starch. Carbohydrate Research. 1985;135(2):271–81.
[32] Miles MJ, Morris VJ, Ring SG. Gelation of amylose. Carbohydrate Research. 1985;135(2):257–69.
[33] Pinto MA, Chalivendra VB, Kim YK, Lewis AF. Effect of surface treatment and Z-axis reinforcement on the interlaminar fracture of jute/epoxy laminated composites. Engineering Fracture Mechanics. 2013;114:104–14.
[34] Glasser WG, De Oliveira W, Kelley SS, Nieh LS. Method of Producing Star-like Polymers from Lignin. Google Patents, 1991.
[35] Kumari S, Chauhan GS, Monga S, Kaushik A, Ahn J-H. New lignin-based polyurethane foam for wastewater treatment. RSC Advances. 2016;6(81):77768–76.
[36] Gupta RK, Ionescu M, Radojcic D, Wan X, Petrovic ZS. Novel renewable polyols based on limonene for rigid polyurethane foams. Journal of Polymers and the Environment. 2014;22(3):304–09.
[37] Shen Y, He J, Xie Z, Zhou X, Fang C, Zhang C. Synthesis and characterization of vegetable oil based polyurethanes with tunable thermomechanical performance. Industrial Crops and Products. 2019;140:111711.
[38] Gholami H, Yeganeh H. Vegetable oil-based polyurethanes as antimicrobial wound dressings: In vitro and in vivo evaluation. Biomedical Materials. 2020;15(4):45001.
[39] Patel MK, Theiß A, Worrell E. Surfactant production and use in Germany: Resource requirements and CO 2 emissions. Resources, Conservation and Recycling. 1999;25(1):61–78.
[40] Stalmans M, Berenbold H, Berna JL, Cavalli L, Dillarstone A, Franke M, et al. European life-cycle inventory for detergent surfactants production. Tenside Surfactants Detergents. 1995;32(2).
[41] Steber J, Herold C-P, Limia JM. Comparative evaluation ofanaerobic biodegradability of hydrocarbons and fatty derivatives currently used as drilling fluids. Chemosphere. 1995;31 (4):3105–18.
[42] Hill K. Fats and oils as oleochemical raw materials. Pure and Applied Chemistry. 2000;72(7):1255–64.
[43] Petrović ZS. Polyurethanes from vegetable oils. Polymer Reviews. 2008;48(1):109–55.
[44] Baber TM, Vu DT, Lira CT. Liquid– liquid equilibrium of the castor oil+ soybean oil+ hexane ternary system. Journal of Chemical and Engineering Data. 2002;47(6):1502–05.
[45] Yeganeh H, Mehdizadeh MR. Synthesis and properties of isocyanate curable millable polyurethane elastomers based on castor oil as a renewable resource polyol. European Polymer Journal. 2004;40 (6):1233–38.
[46] Suresh KI, Thachil ET. Study on blending of millable polyurethane derived from castor oil with natural rubber. Macromolecul Mater Engineer. 1995;224(1):55–60.
[47] Zlatanić A, Lava C, Zhang W, Petrović ZS. Effect of structure on properties of polyols and polyurethanes based on different vegetable oils. Journal of Polymer Science, Part B: Polymer Physics. 2004;42(5):809–19.
[48] Trân NB, Vialle J, Pham QT. Castor oil-based polyurethanes: 1. Structural characterization of castor oil – Nature of intact glycerides and distribution of hydroxyl groups. Polymer (Guildf). 1997;38(10):2467–73.
[49] Ogunniyi DS. Castor oil: A vital industrial raw material. Bioresource Technology. 2006;97(9):1086–91.
[50] Ehrlich A, Smith MK, Patton TC. Castor polyols for urethane foams. Journal of the American Oil Chemists' Society. 1959;36(4):149–54.
[51] Sakamoto WK, Kanda DHF, De Assis Andrade F, Das-Gupta DK. Dielectric relaxation of vegetable-based polyurethane. Journal of Materials Science. 2003;38(7):1465–70.

[52] Fan Q, Xiao C. Effects of crosslinking density on structure and properties of interpenetrating polymer networks from polyurethane and nitroguar gum. Polym Composite. 2008;29(7):758–67.
[53] Ghatge ND, Phadke VB. Elastomers from castor oil. Journal of Applied Polymer Science. 1967;11(5):629–38.
[54] Ping P, Wang W, Zhang P, Chen X, Jing X. Shape-memory and biocompatibility properties of segmented polyurethanes based on poly (L-lactide). Frontiers of Chemistry in China. 2007;2(4):331–36.
[55] Troester G. Method for producing elastomer skins as lining material for plastic molded articles such as automobile dashboards. Google Patents. 1991.

Sheetal Monga, Dheeraj Ahuja, Lokesh Kumar, Anupama Kaushik

4 Polylactide-based shape-memory biodegradable polyurethanes and their potential applications

4.1 Introduction

Poly(lactic acid) or polylactide (PLA) has proven to be the most attractive and useful class of biodegradable aliphatic polyesters in human history. It can be utilized as a replacement for conventional petroleum-based polymers because of its physical, good mechanical, degradation properties, and good processability. Due to various environmental, economic, and safety challenges, biodegradable polymers are gaining importance. PLA as a leading candidate is a thermoplastic and is one of the most promising and extensively researched biopolymers used today. It has numerous applications such as healthcare and medical industry, packaging, automotive applications, and so on.

PLA being a biopolymer exhibits several benefits such as the following [1]:

i) Eco-friendly: It is derived from renewable resources (e.g., corn, wheat, or rice), biodegradable, recyclable, and compostable.
ii) Biocompatible: It is nontoxic. A biocompatible material should not produce toxic or carcinogenic effects in the local tissues. Also, the degradation products should not interfere with tissue healing. PLA hydrolyzes to its constituent α-hydroxy acid when implanted in living organisms, including the human body. Then it is incorporated into the tricarboxylic acid cycle and excreted. Moreover, PLA breaks down into nontoxic by-products during degradation (at a lower composition), making it a natural choice for biomedical applications [2]. The Food and Drug Administration (FDA) has also approved PLA for direct contacting with biological fluids [3].
iii) Processability: It has better thermal processability compared to poly(hydroxyalkanoate) (PHA), poly(ethylene glycol) (PEG), and poly(ε-caprolactone) (PCL). It can be processed by injection molding, film extrusion, blow molding, thermoforming, fiber spinning, and film forming [4].
iv) Energy savings: PLA production requires 25–55% less energy as compared to petroleum-based polymers and estimations show that this can be further reduced to less than 10% in the future. Consumption of lower energy in PLA production is potentially advantageous with respect to cost as well.
v) Bioabsorbable: It can be safely absorbed by the body because the monomers may be produced from nontoxic renewable feedstock as well as due to being a naturally occurring organic acid.

PLA is the first commodity polymer obtained from renewable resources and is recognized as safe by the FDA [5]. It is classified as a safe polymer for all food packaging

https://doi.org/10.1515/9783111145136-004

applications. It is the highest potential biopolymer due to its abundant availability and low cost. Healthcare and medical applications of PLA include manufacturing of delivery system materials, tissue engineering scaffolds, or covering membranes and different bioabsorbable medical implants. PLA has been investigated for membrane applications (e.g., wound covers), implants and medical devices (plates, fixation rods, pins, screws, sutures, etc.), and dermatological treatments (e.g., facial lipoatrophy and scar rejuvenation) due to its versatility. Ongoing years, PLA has gained enormous attention and is utilized at large scale as an alternative to petrochemical-based synthetic polymers (polyethylene terephthlate (PET), polystyrene (PS), polyethylene (PE), etc.) in packaging and textile sectors due to new techniques that allow the economical production of high-molecular-weight PLA [4].

PLA is industrially produced through the polymerization of lactic acid (LA) or by the ring-opening polymerization (ROP) of lactide (the cyclic dimer of LA, as an intermediate) [6]. PLA is produced from nonfossil renewable natural resources by fermentation of polysaccharides or sugar, for example, extracted from corn, potato, cane molasses, sugar-beet, and so on, therefore allowing to complete the biological cycle with PLA biodegradation, as well as the photosynthesis process [7]. Although PLA has many advantages mentioned earlier, there are still some shortcomings associated with the polymer that limit its use in certain applications:

i) PLA's poor ductility, brittleness, and low impact strength limit its use as compared to other thermoplastics such as acrylonitrile butadiene styrene.
ii) Its low crystallization rate and processing results in amorphous products.
iii) PLA is more susceptible to chemical and biological hydrolysis as compared to aromatic polyester (e.g., PET).
iv) PLA has poor gas barrier performance and is thermally unstable.
v) It has low flexibility and is relatively hydrophobic.
vi) PLA has a slow degradation rate. The degradation rate depends on the crystallinity of PLA, weight-average molecular weight (M_w), and its distribution, morphology, stereoisomeric content, and water diffusion rate into the polymer. Its degradation takes place through the hydrolysis of backbone ester groups. For biomedical applications, the degradation rate is considered to be an important parameter. The slow degradation rate might be a serious problem with reference to disposal of consumer commodities and it may lead to a long in vivo lifetime, which could be up to years (3–5 years in some cases) as well [1].
vii) It is chemically inert with no reactive side-chain groups, making its surface and bulk modification a difficult task.

PLA has shape-memory properties and good mechanical strength; however, it does have limitations related to its thermal and physical properties to be used as shape-memory material in human body. It has a melting temperature (T_m) of 170–183 °C and tends to be brittle and its glass transition temperature (T_g) is much higher than body temperature, that is, about 60–65 °C [8] and, thus, cannot be utilized directly in human body as shape-memory material [9]. With improved mechanical strength, physical properties, and enhanced shape-memory properties, PLA-based polyurethanes (PUs) with switching/transition temperature around the body temperature are gaining importance.

PUs are a broad class of polymers having urethane group (–NHCO–O–) in their backbone chain. The repeating unit, that is, urethane linkage is produced from the reaction of an isocyanate (–NCO) with an alcohol (–OH). PUs are one of the most useful commercial classes of polymers that are widely used in both industry and everyday life.

PLA-based PUs (PLAUs) are formed by the polymerization reaction of three basic components: a diisocyanate (NCO–R–NCO), a short-chain diol or chain extender (OH–R'–OH), and a long-chain polylactide diol (OH–PLA–OH). The properties of Pus, to a large extent, depend upon the choice of raw material components such as di-/tri-isocyanates, polyols, as well as chain extenders. The most commonly used polyols are polyethers, polyesters, or acrylic polyols. On the other hand, methylene diphenyl diisocyanate (MDI), hydrogenated MDI (H12MDI), isophorone diisocyanate (IPDI), toluene diisocyanate (TDI), 1,5-naphthalene diisocyanate, hexamethylene diisocyanate (HDI), and xylene diisocyanate are widely used diisocyanates in PU formulations. The structure of thermoplastic segmented PU consists of blocks or segments joined together by covalent bonds, making it a block copolymer [10]. These segments are called "soft segments" and "hard segments." Soft segments are made up of long chains of polyols. They have low density of urethane groups, and thus have low polarity and are flexible and soft at room temperature. On the other hand, hard segments consist of diisocyanate and short-chain extender moieties such as diols and diamines. They are highly polar as they have high density of urethane groups, making these segments rigid and hard at room temperature. Hard segments have high interchain interaction due to hydrogen bonding between the urethane groups.

PU block copolymers are typical shape-memory polymers (SMPs) and these received increasing attention, owing to their adjustable transition temperatures and good biocompatibility along with excellent processability, low cost, and good recovery ability. SMP is generally stimuli-responsive material that can change its shape rapidly from temporary shape to its original shape under an external stimulus [11] such as electricity, heat, alternating magnetic field, microwave radiation, and solvent.

SMPs are usually thermoresponsive, that is, at a critical temperature (called transition temperature or switching temperature), their temporary deformation can be eliminated and their permanent shape can be restored, which can be either T_g or T_m of the materials. These polymers exhibit two separate phases, one of which has a higher thermal transition and is responsible for the permanent shape, while the other has a lower thermal transition and enables the formation and fixation of the temporary shape [9]. SMPs are generally physically cross-linked or chemically cross-linked, but most of them are nonbiodegradable. Biodegradable and nontoxic shape-memory PCL-based PU (PCLU) was prepared by Lendlein and Langer and demonstrated its potential applications in human body [12]. The shape recovery temperature of PCLU can be adjusted near the body temperature, that is, 37–40 °C by adjusting hard-to-soft segment molar ratio and the molecular weight of the PCL diol [13]. It was found that the PCLU products after shape recovery are not too strong to be used as medical devices. To seek a shape-memory material with good mechanical strength and shape recovery temperature near the body temperature, Wang et al. [9] synthesized a series of PLAUs and their shape-memory behavior was also examined. Their T_g was found in the range of 33–53 °C and is controlled by the M_n of PLA diol and hard-to-soft segment ratio. These materials can restore their shapes completely after tensile elongation of 150% or a twofold compression. The results showed that PLAUs display better shape-memory behavior under a small deformation and more rapid response to the thermal stimuli as compared to PCLU.

Thus, PLAU, with a unique combination of toughness, flexibility, biocompatibility, and durability, is a suitable biodegradable material for numerous applications in biomedical, construction, and automotive industries. By controlling the chain length and molar ratio of hard-to-soft segments of PU, T_g and shape-memory temperature can be adjusted to the neighborhood of the body temperature and optimal flexibility and toughness can be tailored to satisfy the needs of various medical applications.

4.2 Polylactide polymers

Generally, LA (2-hydroxy propionic acid) is produced by the fermentation of sugar cane and corn. LA is biodegradable and compostable with low or no toxicity and high mechanical strength comparable to those of commercial polymers. LA is a constituent unit of PLA. Since it is a chiral molecule, it exists as two enantiomers: L- and D-lactic acid. The L-LA is first converted into the cyclic dimer, or can be converted into a mixture of the three stereoisomers, namely, L-lactide, mesolactide, and D-lactide (Fig. 4.1) in the presence of a suitable catalyst. PLA has stereoisomers, namely, poly(L-lactide) (PLLA), poly(D-lactide), and poly(DL-lactide) (PDLLA). Generally, pure L-LA is used for PLA production. The polymerization of racemic (D, L)-lactide and mesolactide yields amorphous polymers. PLA is a crystalline polymer (close to 37% crystallinity) and the degree of crystallinity depends on the molecular weight and polymer processing parameters.

4.2.1 Polyols

High-molecular-weight compounds containing two or more reactive hydroxyl groups synthesized from an initiator and monomeric building blocks are known as "polyols." A wide range of polyols with different chemical structures and hydroxyl numbers are available for PU synthesis. They can be classified as polyester polyols, polyether polyols, dihydroxy and polyhydroxy alcohols, oil-based polyols, bio-based polyols, lactide-

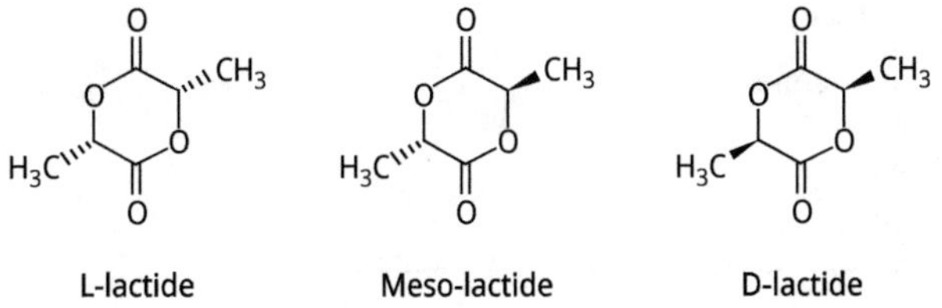

Fig. 4.1: Different isomers of lactide [14].

based polyols, lignin-based polyols, starch-based polyols, and other bio-based polyols. The basic chemical structure and the average number of hydroxyl groups available in polyol provide extreme flexibility to maneuver the formulations for obtaining desired characteristics of the final product. Polyols are generally classified into two main categories, that is, polyester polyols and polyether polyols.

A low-molecular-weight (M_w) PLA by heating LA under vacuum was synthesized in 1932 by Carothers (at DuPont). By ROP of the lactide, high M_w PLA was synthesized. In the last few decades, PLA has been fabricated in multiple ways, but none of them is easy to execute. PLA synthesis requires rigorous control of conditions (temperature, pressure, and pH), the use of catalysts, and long polymerization times [1]. PLA can be prepared by different polymerization processes from LA, including polycondensation, ROP, and by direct methods like azeotopic dehydration and enzymatic polymerization. The PLA nomenclature is full of contradictions in the literature, but generally polymers synthesized from lactide by ROP are known as polylactides and ones derived from LA by polycondensation or other routes are referred to as "poly(lactic acid)." Basically both types are called PLA [15].

4.2.2 Methods of synthesis of polylactides

Generally, there are four methods used for the synthesis of PLA (Fig. 4.2).

4.2.2.1 Condensation polymerization (polycondensation)

Direct condensation of LA involves three stages: first is the removal of the free water, second is oligomer polycondensation, and third is melt condensation of high-molecular-weight PLA. Removal of water is the rate-determining step in the first and third stages. For the second stage, the rate-determining step is the chemical reaction, which depends on the catalyst used. Polycondensation is the least expensive route and it includes solution and melt condensation polymerization. LA is polymerized at reduced pressure and high temperature in the presence of a catalyst in direct polycondensation process. The product polymer has a low molecular weight of a few ten

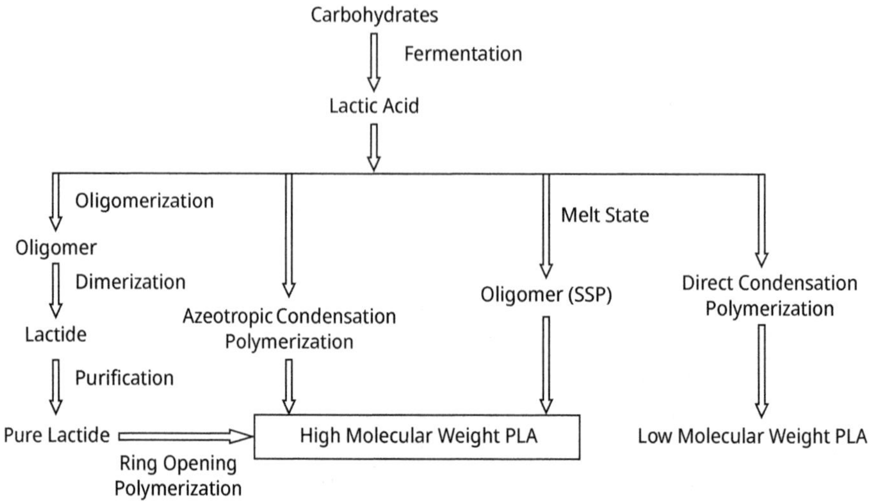

Fig. 4.2: Various methods to synthesize PLA [15].

thousand. It is very difficult to obtain PLA with high molecular weight and high stereoregularity using this process, which restricts its use [4]. The polymer thus obtained has poor mechanical properties. Therefore, this method can be used for producing polymers of low-to-intermediate molecular weight. However, by coupling this material with epoxides, isocyanates, or peroxide, high-molecular-weight polymers can be obtained using this technique [16]. The coupling agent joins the polymer chains of low molecular weight to produce high-molecular-weight polymer. Polymerization of LA in the presence of a small amount of bi-/multifunctional hydroxyl compounds, for example, 2-butene-1,4-diol, 1,4-butanediol (BDO), or glycerol, which results in preferential hydroxyl end groups to PLA, that is, yields hydroxyl-terminated PLA. Similarly, bi-/multifunctional carboxylic acids, such as adipic, succinic, maleic, or itaconic acid, can be used to synthesize carboxyl-terminated PLA, leading to all carboxyl-end functional polymers [17, 18]. Achmad et al. polymerized L-LA by direct polycondensation under vacuum without solvents, catalysts, and initiators at polymerization temperatures of 150–250 °C [19]. The maximum PLA molecular weight obtained was 90 kDa at 200 °C after 89 h, beyond which it degraded thermally. Low-molecular-weight LA polymers were synthesized by condensation polymerization of L-LA by Kari Hiltunen et al. [20]. Depending on the polycondensation conditions and catalyst, the weight-average molecular weights of prepolymers varied from 3,600 to 32,600 g/mol (determined by GPC) and $\bar{M}_n$(NMR) <27,000 g/mol. Direct melt polycondensation of L-LA was performed to produce low molecular weight PLA by Vladimir Sedlarik et al. [21] using methanesulfonic acid (MSA) as a catalyst at different temperatures (130, 145, 160, 175, and 190 °C) and time (6, 12, 18, and 24 h). The results showed that MSA content of

0.5 wt% at 175 °C for 18 h produced PLA with the highest weight-average molecular mass of 17.2×10^3 g/mol.

Polycondensation process may lead to other side reactions such as transesterification that results in the formation of ring structures such as lactide. These side reactions have an adverse effect on properties of the polymer product [22] and can be avoided by using different catalysts, by *controlling the polymerization conditions and functionalization agents.*

4.2.2.2 Solid-state polycondensation (SSP)

As mentioned earlier, the main disadvantages of synthesis of PLA by direct polycondensation technique are low yield and limited molecular weight of the product polymer. Melt/SSP technique helps in achieving higher molecular weight of the PLA polymer as shown in Fig. 4.3. This technique involves four stages. The first three stages are the same as that of direct polycondensation process, namely, removal of free water content, oligomer polycondensation, and melt polycondensation with an additional fourth stage. In the fourth stage, the melt-polycondensated PLA obtained from the third stage is cooled below its T_m with the continuous removal of the by-products either with a carrier, for example, blowing inert gas or by volatilizing under reduced pressure [23]. This step leads to particle formation as it solidifies. Inert gas in the last stage helps removing the condensate from the surface of the material and inhibits polymer oxidation. The solid particles then undergo crystallization. SSP reaction progresses stepwise in the amorphous regions of the semicrystalline polymer, where all the reactive end groups are present. Mass transport of reaction water by molecular diffusion is the rate-determining step in this technique [24]. Moon et al. [23] synthesized PLLA successfully using melt/solid polycondensation of L-LA catalyzed with a tin chloride dihydrate/*p*-toluenesulfonic acid binary system. In this process, water solution of LA is taken as the starting material, and the process is conducted in three steps, namely, dehydration, polycondenastion, and crystallization. In the first step, water is removed by distillation, and an oligomer (oligo(L-lactic acid)) is formed. In the second step, oligomer undergoes polycondenastion in melt state in the presence of a catalyst to obtain low-molecular-weight PLLA prepolymer. In the last step, the prepolymer is then crystallized at 105 °C ($p \approx 70$ Pa) and then it is subjected to SSP carried out at 140–150 °C for 10–30 h under reduced pressure ($p \approx 70$ Pa). PLLA with molecular weight of more than 500,000 Da with high % yield can be obtained in a relatively short reaction time by this technique.

Zhang and Wang [26] used this technique to obtain high-molecular-weight PLA. Various catalysts were used in the first step of melt polycondensation, which is followed by solid polycondensation using *p*-toluenesulfonic acid monohydrate (TSA) as a catalyst. Results indicated that $SnCl_2 \cdot 2H_2O$/maleic anhydride, $SnCl_2 \cdot 2H_2O$/succinic an-

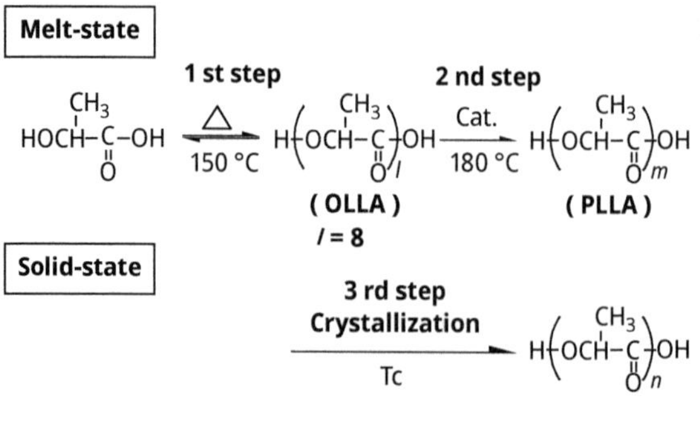

Fig. 4.3: Melt/solid-state polycondensation [25].

hydride, and $SnCl_2 \cdot 2H_2O$/TSA binary catalysts should be very effective to produce PLA with high molecular weight of more than 1.2×10^5 g/mol.

Stereoblock PLA with D- and L-lactate stereosequences was prepared by Fukushima et al. using SSP of a 1:1 mixture of poly(L-lactic acid) and poly(D-lactic acid) [27]. In the first stage, L- and D-LA are subjected to melt polycondensation to synthesize poly(L-lactic acid) and poly(D-lactic acid) with a medium molecular weight, respectively. In the second stage, stereocomplex of poly(L-lactic acid) and poly(D-lactic acid) is formed as these polymers are melt blended in 1:1 ratio. In the third stage, stereocomplex is subjected to SSP at a temperature where chain extension in the amorphous region is promoted with the stereocomplex crystals preserved. A study on polymerization conditions optimization was also conducted. It was concluded that when SSP was performed above 180 °C, the highest M_w of the prepared PLA reached 100,000 Da. The effect of crystallization on SSP of PLLA has also been explored [28]. The prepolymer was crystallized at 105 °C for a range of time intervals and then heated at 135 °C for 15–50 h for further SSP. Experiments revealed that M_w of PLLA reached maximum value for crystallization time of 30 min and SSP time of 35 h.

In direct melt polycondensation of LA, high vacuum and high temperature not only produce dehydration but also favor backbiting reactions such as inter- and intramolecular transesterification, thus preventing the growth of PLA chain due to decomposition of low molecular polymer to lactide. The advantages of SSP include low operating temperatures, which control the side reactions as well as thermal, hydrolytic, and oxidative degradation of the polymer along with reduced discoloration of the product. Monomer cyclization and other side reactions are limited in SSP polymers, leading to the polymers with improved properties. There is practically no environmental pollution involved as no solvent is required in the SSP process [15].

4.2.2.3 Azeotropic polycondensation

It is possible to obtain high-molecular-weight polymer without the use of chain extenders using azeotropic polycondensation technique. In this method, the problem of removing by-produced water is overcome by manipulating the equilibrium ratio of polymer and monomer in an organic solvent. In this approach, LA undergoes reduced pressure distillation for 2–3 h at 130 °C to remove water from polycondensation. Organic solvent diphenyl ether and catalyst are added to LA and then a tube packed with molecular sieves with a size of 3 Å as drying agent is attached to the reaction vessel. The refluxing solvent is returned to the vessel by passing through molecular sieves for an additional 30–40 h at 130 °C. The polymer is then dissolved and precipitated for further purification [29]. Ajioka et al. [30, 31] have synthesized high molecular weight PLA by one-step azeotropic polycondensation of LA in the presence of a catalyst and an organic solvent. The solvent was dried and recycled back into the reaction medium, whereas by-produced water was removed azeotropically. Polymerization temperature was chosen below the melting point of the polymer, thus preventing depolymerization and racemization reactions during the course of polycondensation reaction. Highly pure PLA with molecular weight of more than 300,000 Da was obtained by this method. Kim and Woo [32] reported two different LA polymerization techniques. One was performed through azeotropic dehydration for 48–72 h at 138 °C without using a drying agent. The other polymerization technique was carried out in the presence of a drying agent, that is, molecular sieves of 3 Å size to remove a small amount of water present in the organic solvent. The polymerization was carried out for 30 h using *m*-xylene as a solvent at 138 °C in the presence of 0.2 wt% of $SnCl_2{\cdot}2H_2O$ as a catalyst. The viscosity average molecular weight (M_v) of PLA obtained was about 33,000. While M_v of PLA was a little lower, that is, about 30,000 when polymerization was done in the absence of molecular sieves under the same reaction conditions. Mitsui Toatsu Chemical Company commercialized a process to synthesize poly-DL-lactic acid using direct solution polycondensation. In this process, catalyst, LA, and organic solvent with high boiling point were mixed and azeotropically dehydrated in a reactor for 2 h at 140 °C. The polymer product with a molecular weight of about 300,000 was obtained [33].

4.2.2.4 Ring-opening polymerization

This polymerization technique of LA was first demonstrated in 1932 by Carothers [34] and is represented in Fig. 4.4. High-molecular-weight PLAs were obtained by DuPont in 1954 using improved lactide purification techniques [35]. This method is usually employed for the synthesis of high-molecular-weight polymer with a high degree of stereoregulation. In this technique, PLA is synthesized by the polymerization of their respective cyclic dimmers, that is, LA, which is a six-membered cyclic dimer with a chemical name as 3,6-dimethyl 1,4-dioxane 2,5-dione. LA should be of high-purity grade in this polymerization process because impurities, such as water and oligomers, can interfere with the reaction, leading to the formation of low-molecular-weight polymer with high degree of racemization. The ROP of LAs may be classified as anionic polymerization, cationic polymerization, and coordination–insertion mechanism by their different reaction mechanisms and initiator types [36]:

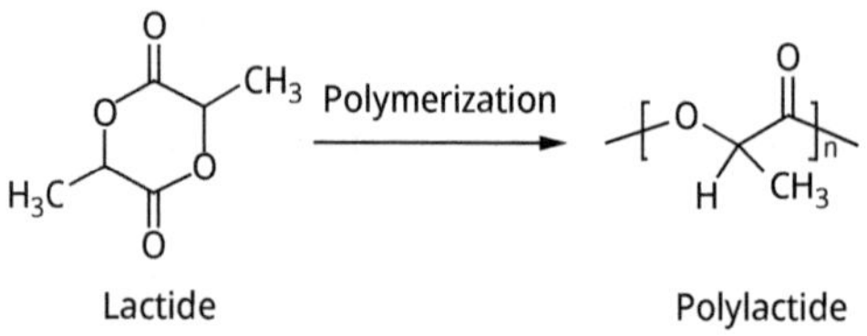

Fig. 4.4: PLA synthesis by ring-opening polymerization.

i) Anionic polymerization: The first step of polymerization, that is, initiation of anionic ROP, is done by alkali metal alkoxide initiator when a carbonyl group of the lactide is attacked by the nucleophilic anion of the initiator, which results in the cleavage of the endocyclic oxygen and carbonyl carbon bond (Fig. 4.5) [15]. The next step is chain growth, which proceeds by alkoxide reactive species. This produces an anion with the alkoxide and carbonyl end group [37], which continues to propagate but nucleophilic initiators are very basic and they deprotonate the monomer resulting in racemization [38].

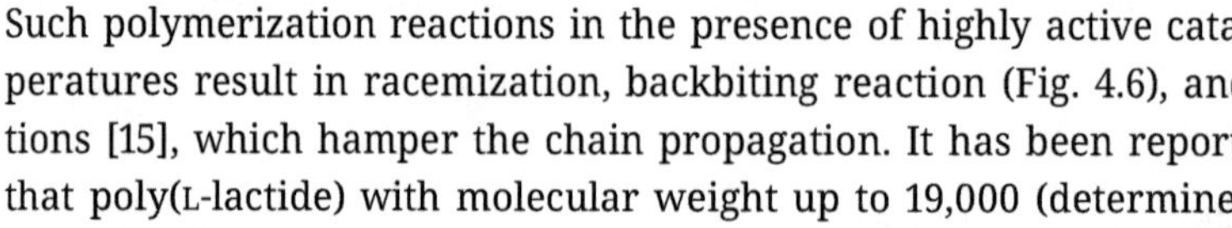

Fig. 4.5: Anionic ring-opening polymerization [15].

Such polymerization reactions in the presence of highly active catalysts at high temperatures result in racemization, backbiting reaction (Fig. 4.6), and other side reactions [15], which hamper the chain propagation. It has been reported by Yaun et al. that poly(L-lactide) with molecular weight up to 19,000 (determined by GPC) can be synthesized using anionic aryloxo cobalt(II) complex, which could effectively initiate the ROP of L-LA both in solution and in bulk [39]. Anionic ring-opening copolymerization of L-LA with a five-membered cyclic carbonate having glucopyranoside structure was studied by Haba et al. [40] using potassium tert-butoxide as initiator and tetrahydrofuran as a solvent at 30 °C for 12 h. The polymer with highest M_n of 12,100 Da was obtained (determined by SEC eluted with DMF using polystyrene standard).

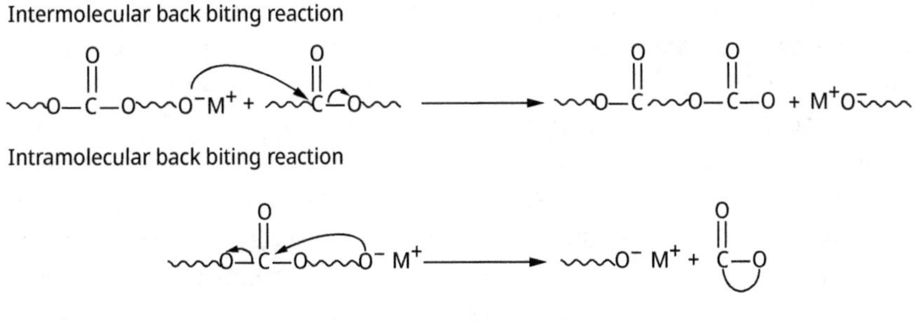

Fig. 4.6: Inter- and intramolecular backbiting reactions [15].

The main disadvantage of using anionic ROP technique is the occurrence of extensive backbiting reactions, and in some cases polyesters with low molecular weight are achieved.

ii) Cationic polymerization: The scheme of cationic ROP is explained in Fig. 4.7. It involves the alkylation or protonation of the carbonyl O-atom that results in electrophilic activation of the O–CH bond. This bond is further cleaved by nucleophilic attack of another monomer to create another electrophilic carbenium ion [41]. This bond again undergoes nucleophilic attack by additional monomers and this process propagates until the polymerization is terminated. High temperature causes racemization in cationic polymerization as the second monomer attacks at the chiral center of the propagating chain. Nonetheless, racemization can be controlled if polymerization is carried out at low temperatures, but the rate of reaction becomes very low at low temperatures and it is difficult to obtain polymers with high molecular weight at low reaction temperatures [37].

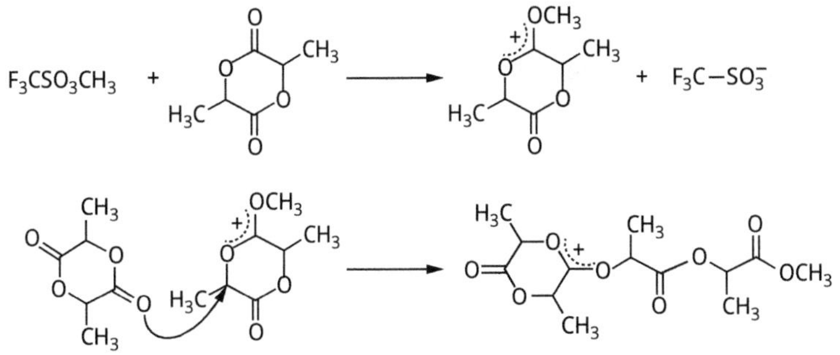

Fig. 4.7: Cationic ring-opening polymerization [15].

iii) Coordination–insertion ROP: The pseudo-anionic ROP is often known as "coordination–insertion ROP." This mechanism involves an initial step of opening the cyclic ring monomer to be polymerized. The propagation step is coordination of the monomer to the active species, which is Lewis-acidic metal center and then insertion of the monomer into the metal–oxygen bond by rearrangement of the electrons [42]. Figure 4.8 shows the mechanism of ROP. In this mechanism, catalysts used are generally metal alkoxides or esters having a covalent metal–oxygen bond and the character of weak Lewis acids [42, 43]. The most commonly used tin initiator is stannous octoate or stannous(II) ethylhexanoate $(SnOct)_2$ [44]. Tin alkoxide is the initiating species of this mechanism, which is formed prior to polymerization. Initiation of coordination–insertion mechanism takes place when one of the exocyclic oxygens of the LA becomes coordinated with the metal atom of the initiator. This coordination enhances the nucleophilicity of the alkoxide part of the initiator along with electrophilicity of the lactide carbonyl group. In the propagation step, the acyl–oxygen bond (between the carbonyl group and the endocyclic oxygen) of the LA is cleaved and the LA chain produced is inserted into the metal–oxygen bond of the initiator. The propagation of LA continues as additional cleaved LA molecules are inserted into the metal and its adjacent oxygen bond. However, the other end of the propagating chain, that is, the alkoxide end of the initiator, becomes a dead chain end. This coordination–insertion mechanism of LA using tin alkoxide as a catalyst is highly preferred for the production of high-molecular-weight polymers, good reaction rates, and low level of racemization [45, 46].

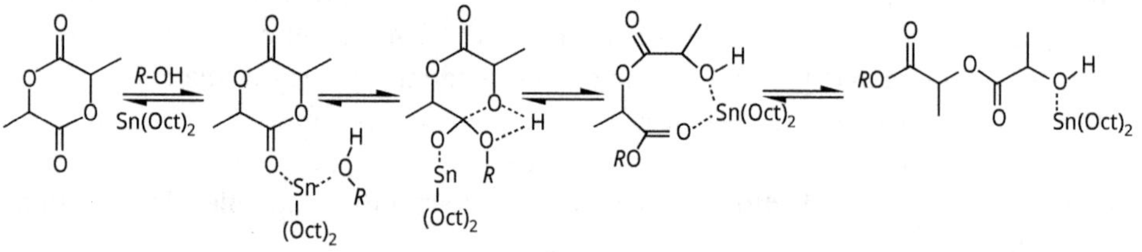

Fig. 4.8: Coordination–insertion ring-opening polymerization [47].

In the light of current development, the advantages and disadvantages of different methods for synthesis of PLA are summarized in Tab. 4.1.

Tab. 4.1: Various polymer synthesis methods with their advantages and disadvantages.

Method	Initiation	Advantages	Disadvantages
Condensation polymerization	Three stages: removal of the free water, oligomer polycondensation, and melt condensation	Least expensive	Difficult to obtain a solvent-free high–molecular-weight PLA [48]

Tab. 4.1 (continued)

Method	Initiation	Advantages	Disadvantages
			Low-molecular-weight PLA of a few 10,000 Low stereoregularity Poor mechanical properties Transesterification High operating temperatures
Solid-state polycondensation	Four stages: removal of the free water, oligomer polycondensation, melt condensation, andcontinuous removal of the by-products either with a carrier or by reducing pressure	Higher molecular weight of the PLA Low operating temperatures No environmental pollution High purity (suppression of side reactions)	Low % yield Long duration Complicated operation
Azeotropic polycondensation	By-produced water is removed azeotropically	High-molecular-weight polymer Low cost Basic equipment Moderate temperature (<180 °C) [49] Prevents depolymerization and racemization reactions	Low yield Low purity (usually with residual solvent and by-products in polymer) Solvent waste and environmental pollution
Ring-opening polymerization	Cyclic monomers initiated by different ions	High-molecular-weight PLAs High degree of stereoregulation Controlled polymer properties Wide range of molecular weight [50]	Low overall yield Long operating time Demanding condition Complicated operation
Anionic polymerization	Initiation is done by alkali metal alkoxide initiator		Racemization Highly active catalyst is needed at high temperature Side reactions

Tab. 4.1 (continued)

Method	Initiation	Advantages	Disadvantages
Cationic polymerization	Electrophilic activation		High temperature causes racemization
Coordination–insertion ROP	Coordination of the monomer to the active species	Good reaction rates Low level of racemization	

4.2.3 Applications of polylactides

PLA is being widely utilized and has potential to be used in multiple applications:

i) Food packaging: PLA is an FDA-approved polymer to be used as a food contact material. It has been used as a food packaging polymer for short shelf-life food items such as fruits and vegetables. PLA packaging is ideal for fresh food stuff and those whose quality is not deteriorated by PLA oxygen permeability. PLA is being considered as a "green" food packaging polymer. Thermoformed PLA containers are used in retail markets for fruits, vegetables, and salads. PLA imparts functional attributes like stiffness, clarity, dead fold and twist retention, low-temperature heat sealability, and special aroma and flavor barrier characteristics to the packaging. The common food-contact articles made up of PLA include drinking cups, plates, salad cups, disposable cutlery, overwrap and lamination film, stirrers, straws, blister packaging, lids and cups, and containers for food dispensed at delicatessen and fast-food establishments [5].

ii) Fiber and fabric application: PLA can be designed for various fiber applications as it is readily melt spinnable and stress crystallizes upon drawing. PLA as fiber has found potential applications as hollow fiberfill for pillows and comforters, filament yarns, bulk continuous filament for carpets, and spun yarns for apparel, spun bond, and other nonwovens and bicomponent fibers for binders and self-crimping fibers. PLA has been gaining importance as a nonwoven textile to be utilized as upholstery, awnings, disposable garments, feminine hygiene products, and nappies [1]. PLA as a fabric material has been used in making carpets, shirts, mattresses, bedding, sportswear, and so on due to its low smoke generation capability, low moisture absorption as well as good resistance to UV light. It also has potential in mulch films and compostable garden waste bags, insulation materials, structural protective foams, and so on.

iii) Healthcare and medical industry: Due to unique attributes of PLA such as biodegradability, biocompatibility, thermoplastic processability, and eco-friendliness, it has been utilized extensively in healthcare and biomedical fields. It is used in fabricating tissue engineering scaffolds, delivery system materials, or covering membranes and different bioabsorbable medical implants. Being a versatile material, it has been investigated for membrane applications (e.g., wound covers), implants and medical devices (e.g., surgical sutures, stents, plates, and screws for craniomaxillofacial bone fixation; interference screws in ankle, knee, and hands; tacks and pins for ligament attachment, spinal cages, anchors, soft-tissue implants, etc.), and dermatological treatments (e.g., facial lipoatrophy and scar rejuvenation).

iv) Structural applications: Bio-based PLA and its composites show excellent tensile strength and impact characteristics; therefore, these are recently being used for high-end automotive as well as electrical and electronics applications. These composites could be used for safety helmets and interior parts of automobiles. PLA composites with enhanced material properties meet all the requirements to be used in pillar cover, front panel, floor mats, door trim, and ceiling material.

4.3 Polylactide-based polyurethanes

4.3.1 Polyurethane structure

PUs are the class of polymers that are composed of organic units joined by carbamate or urethane links (–NHCOO–) [51] and were first produced by Bayer in 1937. These consist of a two-phase structure, a hard segment and a soft segment, and are generally obtained by condensation reaction of diisocyanates and polyol compounds [52–54]. PU polymers have segmented structures where hard and soft segments alternate along the chain. Hard segment domains are semicrystalline and rigid that appear as dispersed in the matrix of crystallizable soft segments (Fig. 4.9) [60]. Segmented PUs have a special behavior of phase separation because of incompatibility between the soft and hard segments. Hard segment clusters are generally of the dimensions <1 μm and possess higher T_g and T_m values than the ones of soft segment and act as cross-links in the matrix. A soft segment domain is amorphous in nature with low T_g values. The hard segments are mainly composed of the aromatic or aliphatic diisocyanate and the chain extender. Aromatic isocyanates are more reactive than the aliphatic varieties. Chain extenders may be low-molecular-weight diols or diamines used to further couple these prepolymers. The soft segments are derived from macroglycol moieties that are used to maintain the temporary shape of the polymer [55]. Soft segments are generally composed of polyester, polycarbonate, or polyether macroglycols [56, 57]. High-molecular-weight polyols, which form soft segments, provide rubbery properties to PU. The hard segments, which are composed of isocyanate and low-molecular-weight diols, provide multifunctional physical net points and crystalline chains for PU. Hard segments have high density of urethane groups of high polarity and are rigid at room temperature. Soft segments are flexible at room temperature because they have low density of urethane groups and thus have low polarity. When thermodynamically incompatible hard and soft segments are joined end-to-end via covalent urethane bonds, these give rise to multiblock PU copolymers. Due to difference in polarities of hard and soft segments, they do not mix on a molecular level, resulting in a microphase-separated structure of PU with unique properties. The phase separation of the hard and soft segments is responsible for the elastomeric properties of these polymers as urethane hard segment domains serve as cross-links between the amorphous soft segment domains. As the concentration of the soft segment increases, the resultant material becomes flexible and elastomeric. The property of segmented PU is the added sum of the hard segments and the soft segments. The extent of phase separation depends mainly on the length of the hard and soft segments. If the soft segments are too long, then material will be phase separated, and if they are too short, then the possibility of phase separation is decreased [58].

PUs have been found very useful in various industrial applications [59] such as sealants, coatings, adhesives, and textile dyes. PUs have also gained importance in biomedical field to be used as stent, sutures, drug delivery devices, and orthopedics due to their durability, elasticity, biostability, fatigue resistance, biodegradability, compliance, and tailorable backbone structure from a wide range of available precursors.

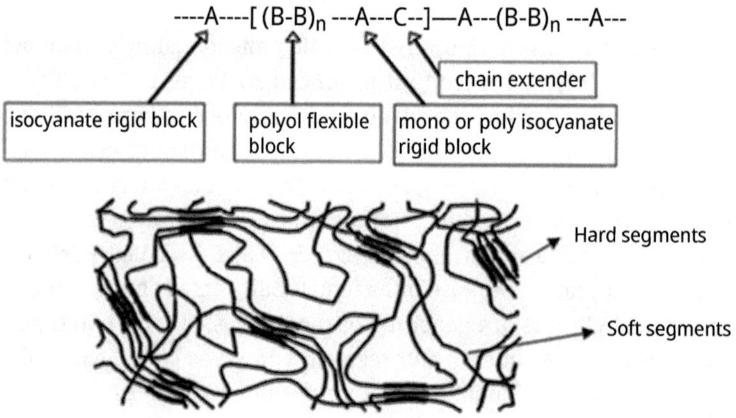

Fig. 4.9: Polyurethane structure representing hard segment dispersed in soft crystallizable phase [60].

4.3.2 Biodegradable polylactides and polyurethanes

Bio-based and biodegradable are two important and independent characteristics of a polymer. Biodegradable PUs can be degraded to noncytotoxic products in vivo and in vitro by hydrolytic or enzymatic decomposition. Hydrolysis of ester linkages in PU structure yields hydroxyl acids and urethane/urea fragments with terminal acid groups as degradation products. The composition of the soft segment (i.e., polyester polyol component) of the polymer controls the degradation rate in vitro. Additionally,

PU elastomers having amorphous soft segments have been discerned to degrade more rapidly than polymers with semicrystalline soft segments. PUs with hydrophilic soft segments undergo rapid degradation due to increased water uptake. Thus, the hydrolysis process is considered to be the primary mechanism for the degradation of PUs. Water absorption is a key parameter to evaluate degradation of polymer samples by water sorption due to its more amorphous morphology and higher free volume. A hard segment provides tortuous path for the diffusion process. There is a direct correlation between the water sorption of polymers and their degradation profile, which can be assessed by molecular weight loss and weight loss. Polymers having hydrophilic groups or higher water sorption properties show a speedy degradation rate, as the water molecules can easily penetrate into the internal structure of the polymer chains that accelerate the hydrolytic degradation. Degradation products could be retained in the bulk polymer structure hydrogen bonding, van der Waals force, polar interaction, and so on. These polymers contain physical cross-linking between the hard segments; therefore, they keep the polymer films intact after degradation, while sharp decrease in molecular weight can be observed during in vitro degradation. Since hydrolytic degradation takes place preferably in the amorphous region, the water molecules can easily diffuse into the amorphous region of the polymer.

The hydrolysis process in PUs occurs mainly in two steps. In the first step, water molecules attack ester and urethane groups in the polymer bulk and the long polymer chains are broken into short polymer chains releasing acidic substances. At low pH, hydrolysis is self-accelerated in bulk, resulting in drop in molecular weight of PU but the mass loss lags behind it because those degraded polymer chains are still water insoluble. In the second step, the short polymer chains are further hydrolyzed into small water-soluble molecules, which is further responsible for drop in PU molecular weight, decrease in pH, and mass loss.

In this area, PCLUs, polycaprolactone/polylactide blends, PLA polyester-based PUs, poly(butylene succinate) (PBSU), and PHA have been studied extensively due to their two major fetching advantages, that is, independence from petroleum and complete biodegradation into nontoxic degradation products. Here "biodegradation" means hydrolyzable at temperatures up to 50 °C (e.g., in composting) over a period of several months to one year with the help of naturally occurring microorganisms such as bacteria, algae, and fungi [15]. Generally, a biodegradable polymer is "a degradable polymer wherein the primary degradation mechanism is through the action of metabolism by microorganisms." Biodegradation in this context is considered a category of degradation involving biological activity. The chemical structure of the polymer, which is responsible for reactivity, functional group stability, swelling behavior, and hydrophilicity, is the key factor for determining the biodegradability of polymers. Other important factors responsible for biodegradation of polymeric materials are physical and physicomechanical properties, for example, molecular weight, elasticity, porosity, and crystalline/amorphous nature [61]. This process refers to the degradation and assimilation of polymers by living microorganisms to produce nontoxic degradation products, which is one of the important prerequirements for any potential biomedical applications:

a. Degradation of polylactides: The polyester and copolyesters of several α-, β-, and ω-hydroxy acids have been investigated and utilized widely during the past 20 years. Makino and coworkers [62] used five buffer solutions having pH values 1.6, 3.0, 5.0, 7.4, and 9.6 to analyze the effect of pH medium on the degradation rate of PLA microcapsules. Fall in M_w of PLA in a degradation medium was a clear indication of cleavage of ester bonds in the polymer structure. It was also observed that the degradation of polymer was faster in highly alkaline buffer solutions (pH 9.6) than in the highly acidic and slightly alkaline ones. In vitro degradation of electrospun polyglycolide (PGA), PLA, and poly(lactide-*co*-glycolide) (PLGA) nanofiber matrices was examined by Young You et al. [63] in phosphate buffer solutions of pH 7.4 at 37 °C. The degradation rate of the nanofiber matrices was found to be rapid that follows the order of PGA > PLGA ≫ PLA. Analysis of PGA matrix by wide-angle X-ray diffraction disseminated a significant increase in the crystallinity during the early stage and a gradual decrease during the later period. This indicated the occurrence of preferential hydrolytic degradation in the amorphous regions with cleavage-induced crystallization, followed by further degradation in the crystalline region. In vitro degradation behavior of PLA and PLGA films in Hank's solution was investigated by Wei et al. [64]. The hydrolytic degradation of PLA and PLGA was monitored by measuring the changes of inherent viscosity and weight loss of the resulting samples. The degradation rate of PLGA could be adjusted by varying the composition of the copolymer. An increase in degradation rate was observed with an increase in glycolide content. PLGA copolymer could find its potential application as drug elusion coating on coronary stents. Riitta Suuronen et al. incubated self-reinforced PLLA plates in vitro and in vivo for 5 years [65]. The material degraded considerably faster in vivo than in vitro. The material was almost completely resorbed, but macroscopic particles of the polymer could be noticed at the implantation site after 5 years of in vivo follow-up. Loss in mass of the plates was observed to be 52% ± 8% after 5 years of incubation in vitro.

These degradable polyesters have found potential applications in the development of implanted medical devices for dental, craniomaxillofacial, orthopedic, hernioplastic, and skin surgery. Many medical devices like biodegradable screws, bioresorbable surgical sutures, biodegradable membranes for periodontal treatment, plates for cartilage and bone fixation, surgical meshes, and wound coverings have also been developed [66].

b. Degradation of PUs: A number of polyester-based PUs with desired properties like biodegradation, biocompatibility, and so on have been reported in literature to explore their potential application as biomaterials.

Trinca and Felisberti [67] synthesized biodegradable binary and ternary segmented PUs based on low molar mass (2 kDa) homopolymers-PEG, PLA and poly(trimethylene carbonate) (PTMC), and TDI using 1,4-BDO as a chain extender and dibutyltin dilaurate as a catalyst. Water swelling experiments were performed to analyze the hydrophilicity of the segmented PUs by measuring the swollen mass of water at 25 and 37 °C after 48 h of thermal and water sorption equilibrium. The results showed that precursors have different affinities for water, that is, PEG is a hydrophilic and water-soluble polymer and PLLA and PTMC are hydrophobic and water-insoluble polymers. PLLA- and PTMC-based PUs show very little water uptake (maximum 2%), while with an increase in relative amount of PEG increases the water uptake of the PUs (water uptake of almost 150%) resulting in a hydrogel. PLLA blocks are responsible for preserving the dimensional stability of swollen films. It was observed that with an increase in temperature from 25 to 37 °C resulted in a decrease in the water uptake for PEG-rich PEG-PLLA and PEG-PTMC PUs (enthalpic control). A tendency of increasing water uptake in the case of ternary PUs containing 25 wt% PEG with an increase in temperature is observed (entropic control).

Hydrolytic degradation behavior of poly[(rac-lactide)-co-glycolide]-based polyesterurethanes (PLGA-PU) was studied by Ghobadi et al. [68]. It was observed that PLGA-PU matrices have higher water uptake as compared to pure PLGA. The urethane chain segments have higher mobility, which resulted in an accelerated hydrolytic chain scission rate of PLGA-PU in comparison to pure PLGA.

Cross-linked PUs with 10 and 30 wt% amount of synthetic poly([*R*,*S*]-3-hydroxybutyrate) (*R*,*S*-PHB) in soft segments) and their physical blends with PDLLA were investigated by Brzeska et al. [69] and then degraded under hydrolytic (phosphate buffer solution) and oxidative ($CoCl_2/H_2O_2$) conditions. Degradation studies revealed the small weight loss in samples, change in thermal properties of polymers, and topography of their surface, which suggested that they are susceptible to gradual degradation under oxidative and hydrolytic conditions. Blends of PDLLA and PU with 30 wt% of *R*, *S*-PHB in soft segments and PUR/PDLLA blends absorbed more water and degraded faster than PU with low amount of *R*,*S*-PHB.

Kucharczyk et al. [70] conducted studies on the degradation of PUs based on poly (lactic acid)/poly(ethylene glycol), containing three differing concentrations of diisocyanate compound, under three conditions (hydrolytic, thermal, and composting). Biodegradation process initiated with low-molecular-weight fractions being mineralized by microorganisms. The presence of hydrophilic PEG could be accountable for the fast decomposition, which effectively increased the amount of water uptake into the polymer matrix, triggering accelerated depolymerization of chains by the hydrolysis of ester bonds. Finally, the diffusion rate of H_2O, promoting hydrolytic depolymerization, is higher in the disorganized amorphous phase. A biodegradation experiment showed that all the investigated samples successfully (>80%) mineralized within 90 days under composting conditions.

As discussed earlier, PUs synthesized by a reaction between diols and diisocyanates result in hydrogen-bonded urethane linkages with good physical and mechanical properties and with the haemocompatibility and flexibility required for biomedical applications. Biodegradable segmented PUs with low in vitro and in vivo cytotoxicity have been used as blood compatible materials and tissue-engineering scaffolds [71]. Biodegradable, biocompatible polyester-based PUs with good mechanical properties have been employed in soft tissue reconstruction, such as bladder, tracheal, and cardiovascular applications [72].

4.3.3 Literature on polylactide-based shape-memory polyurethanes

Nontoxic and biodegradable SMPs have become a new category of intelligent polymer materials to be used in biomedical area and found its potential applications in human body. PLAUs are synthesized by reacting PLA diols with diisocyanates in the presence of chain extender. Shape recovery temperature of PU can be adjusted near the body temperature, that is, 37–40 °C by adjusting the molar ratio of the soft-to-hard segment and the molecular weight of the diol. Shape-memory PUs should possess high restoration force and high mechanical strength to be used as medical implants in human body.

Wang et al. investigated a series of PLAUs synthesized from PLA diols, HDI, and 1,4-BDO [9]. These materials possess good shape recovery, that is, can restore their shapes almost completely after 150% elongation or twofold compression. The T_g was found in the range of 33–53 °C. The effect of various hard segments (MDI, TDI, and IPDI) on properties of PLAUs was also studied by Wang et al. [73] using PLA-diol as a soft segment and 1,4-BDO as a chain extender. It was found that MDI-based PLAU has the highest T_g, maximum restoration force and tensile strength, the TDI-based PLAU has the lowest T_g, while the IPDI-based PLAU has the highest tensile modulus and elongation at break. These amorphous PLAUs could restore their shapes almost completely with a tensile elongation of 150% or a twofold compression. Their temporary shape can be maintained easily at room temperature (20 °C). Moreover, they deformed and recovered at a temperature below their T_g values. Peng Ping et al. [74] reported on shape-memory and biocompatible PLAUs synthesized with oligo-poly(L-lactide) as the soft segments and the reaction product of TDI and ethylene glycol (EG) as the hard segments. It was found that the 50% compressed PLAUs could recover their original shape almost 100% within 10 °C from the lowest recovery temperature (22–37 °C). In the recovery process, the PLAU showed a maximum contracting stress in the range of 1.5–4 MPa. Cell incubation experiments showed that PLAU has biocompatibility comparable to that of pure PLA.

Completely amorphous PLA poly(ester urethane) elastomer that exhibits excellent shape fixity (>99%) and shape recovery (>99%) in a time frame of seconds is reported by Fan et al. [75]. By manipulating stereoisomeric ratios and control over architecture, the resultant poly(PLLA/PDLLA urethane)s (PLDU) elastomers showed a single T_g, as the only thermal event, in the range of 38–46 °C. The elastic moduli of PLDU elastomers displayed a 100-fold loss during the sharp transition from a glassy to rubbery state with temperature alteration across their corresponding T_g, indicating a successful manipulation of the thermomechanical properties by temperature as required in thermally induced SMPs.

4.3.4 Polylactide-based shape-memory polyurethane nanocomposites

In last few years, researchers have been engaged in designing and manufacturing environment friendly and sustainable polymeric materials that are primarily derived from biodegradable polymers (either natural or fossil fuel) with enhanced properties. With the aim to achieve desirable high-performance and lightweight green polymeric products, the focus has been given on the use of nano-reinforcements in the polymer matrices. A unique category of composites has evolved in which nanofillers including clays, titanium oxide, carbon nanotubes, silica, nanocalcium carbonate, silver, hydroxyapatite, and nanocellulose crystals (at least one dimension within the range of 1–100 nm) are incorporated and recognized as bionanocomposites [76].

Polylactide-based PU shape-memory nanocomposites (Fe_3O_4/PLAUs) with fast magnetic responsiveness were reported by Gu et al. [77]. Nanocomposites with low Fe_3O_4 loading content showed homogeneous distribution of Fe_3O_4 nanoparticles in the polymer matrix, while nanocomposites with 6 and 9 wt% nanoparticles loading content showed agglomeration leading to a small decline in the mechanical properties. PLAU and its nanocomposites have T_g around 52 °C and shape fixity ratios above 99%, shape recovery ratios above 91% for the second cycle. These polymers also exhibit a fast water bath or magnetic responsiveness. Nanocomposites have fast responses in an alternating magnetic field and have potential application in biomedical areas such as intravascular stent.

Zia et al. [78] prepared halloysite nanotubes (HNTs)/PU bionanocomposites using 2-hydroxyethyl cellulose graft polylactic acid copolymer (HLAC) as a chain extender in the step-growth polymerization. Lower contents of HNTs were completely exfoliated in the HLAC/PU matrix. The crystalline pattern of the elastomers improved at lower contents of HNTs that enhanced the thermal stability of the bionanocomposites. The mechanical testing suggested that HNTs/HLAC/PU bionanocomposites have higher values of tensile strength and % elongation with only 0.3–0.5 wt% contents of HNTs. The amphiphilicity, biocompatibility, and biodegradability of HLAC, the hydrophilic and encapsulating ability of native HNTs along with good mechanical and thermal properties of PU make these bionanocomposites elastomers a potential biomaterial to be used for various applications.

Biodegradable shape-memory poly(D, L-lactide-co-ε-caprolactone)-based PU nanocomposites (PCLAU/Fe_3O_4) were developed for self-expandable stent applications by Gu [79]. The stent could maintain its temporary shape at body temperature for a certain period of time and recover to its original shape at a temperature little above body temperature (around 40 °C) in both a water bath and an alternating magnetic field. The nanocomposites had high fixing ratios above 99% and recovery ratios above 82% at both 37 and 40 °C. Cytotoxicity and in vitro degradation showed the nanocomposites had good biocompatibility and biodegradability. The PCLAU/Fe_3O_4 nanocomposites with dual-responsive shape-memory effects, biocompatibility, desirable mechanical properties, and biodegradability show great potential for vascular stents.

4.4 Conclusions

PU is a leading member of the wide-ranging and highly diverse family of polymers or plastics that are widely used in both industry and in everyday life. PUs are formed by the polymerization reaction of three basic components: diisocyanate, short-chain diol or chain extender, and long-chain polyol. Common applications of PU include adhesives, coatings, automotive, footwear, furniture and bedding, refrigerators, building insulation, biomedical devices, and implants. In this research work, PUs with shape-memory property that have potential applications in biomedical field have been extensively explored. In this area, PCL-based PUs, PCL/PLA blends, polyester-based PUs from PLA, PBSU, and PHA have been studied because of their biodegradability, independence from petroleum, and nontoxic degradation products. SMPs are a class of intelligent materials and they have been widely used in the past decades. Most of the SMPs are thermally sensitive, that is, they respond to thermal heating (hot liquid or gas) to recover their original shape by releasing the mechanical energy exerted on the polymer during deformation. Many researchers have synthesized biodegradable SMPs and demonstrated their potential applications in biomedical field, which are used as drug delivery systems, stents, and implants. Present research work focuses on biodegradable PLAUs with shape-memory properties. PLA, being biodegradable polyester produced from renewable resources, is used for various applications such as biomedical, packaging, textile fibers, and technical items. PLA has a key position in the market of biopolymers due to its inherent properties and it is one of the most promising candidates for further developments. PLA is eco-friendly, biocompatible, bioabsorbable, obtained from renewable resources, and is recognized as safe by the FDA. PLA has good mechanical strength and shape-memory properties; however, PLA does have limitations related to its physical and thermal properties to be used in human body for medical applications as shape-memory material. It has T_m of 170–183 °C and tends to be brittle and its T_g is about 60–65 °C, much higher than body temperature and hence cannot be used directly in human body as shape-memory material. With the intention of seeking a shape-memory material with enough mechanical strength and a switching temperature around the body temperature, researchers have explored a series of shape-memory PLAUs. Most of the SMPs are thermoresponsive; however, it is not suitable enough to acquire the shape-memory effect by direct heating in human body. Thus, inductive heating will be a solution. Plethora of research has been done to realize the shape-memory effect by heating from inside particles of SMPs including carbon nanotubes, carbon nanofibers, and magnetic particles. Considering the potential biomedical applications, these SMP nanocomposites should meet some basic requirements like the materials should be safe and nontoxic, the shape recovery temperature must be slightly higher than body temperature, and the shape recovery rate must be high. Biodegradable shape-memory PLAUs have been utilized as vascular stents, tissue engineering, smart punctual plugs, and drug-loading systems.

References

[1] Farah S, Anderson DG, Langer R. Physical and mechanical properties of PLA, and their functions in widespread applications-A comprehensive review. Advanced Drug Delivery Reviews. 2016;107:367–92.

[2] Rasal RM, Janorkar AV, Hirt DE. Poly(lactic acid) modifications. Progress in Polymer Science. 2010;35 (3):338–56.

[3] Gupta B, Revagade N, Hilborn J. Poly(lactic acid) fiber: An overview. Progress in Polymer Science. 2007;32(4):455–82.

[4] Auras R, Harte B, Selke S. An overview of polylactides as packaging materials. Macromolecular Bioscience. 2004;4(9):835–64.

[5] Jamshidian M, Tehrany EA, Imran M, Jacquot M, Desobry S. Poly-lactic acid: Production, applications, nanocomposites, and release studies. Comprehensive Reviews in Food Science and Food Safety. 2010;9(5):552–71.

[6] Vink ETH, Davies S. Life cycle inventory and impact assessment data for 2014 Ingeo polylactide production. Industrial Biotechnology. 2015;11(3):167–80.

[7] Murariu M, Dubois P. PLA composites: From production to properties. Advanced Drug Delivery Reviews. 2016;107:17–46.

[8] Domb AJ, Khan W. Focal Controlled Drug Delivery. New York, NY: Springer, 2014.

[9] Wang W, Ping P, Chen X, Jing X. Polylactide-based polyurethane and its shape-memory behavior. European Polymer Journal. 2006;42(6):1240–49.

[10] Monga S, Kaushik A, Gupta B. Synthesis of L-Lactide-Based Segmented Polyurethanes. Polymer-Plastics Technology and Engineering. 2016;55(9):943–48.

[11] Qian W, Song Y, Shi D, Dong W, Wang X, Zhang H. Photothermal-triggered shape memory polymer prepared by cross-linking porphyrin-loaded micellar particles. Materials. 2019;12(3):496.

[12] Lendlein A, Langer R. Biodegradable, Elastic Shape-Memory Polymers for Potential Biomedical Applications. Science. 2002;296(5573):1673–76.

[13] Ping P, Wang W, Chen X, Jing X. Poly (ε-caprolactone) polyurethane and its shape-memory property. Biomacromolecules. 2005;6(2):587–92.

[14] Blackburn RS. Biodegradable and Sustainable Fibres. London: Taylor & Francis US, 2005.

[15] Gupta AP, Kumar V. New emerging trends in synthetic biodegradable polymers-Polylactide: A critique. European Polymer Journal. 2007;43(10):4053–74.

[16] Sarin R, Arora AK, Ranjan R, Gupta AA, Malhotra RK. Bio-diesel lubricity: Correlation study with residual acidity. Lubrication Science. 2007;19(2):151–57.
[17] Bonsignore PV. Production of high molecular weight polylactic acid. Google Patents, 1995.
[18] Ibay AC, Tenney LP. Polymers from hydroxy acids and polycarboxylic acids. Google Patents, 1993.
[19] Achmad F, Yamane K, Quan S, Kokugan T. Synthesis of polylactic acid by direct polycondensation under vacuum without catalysts, solvents and initiators. Chemical Engineering Journal. 2009;151(1–3):342–50.
[20] Hiltunen K, Harkonen M, Seppala JV, Vaananen T. Synthesis and Characterization of Lactic Acid Based Telechelic Prepolymers. Macromolecules. 1996;29(27):8677–82.
[21] Sedlarik V, Kucharczyk P, Kasparkova V, Drbohlav J, Salakova A, Saha P. Optimization of the reaction conditions and characterization of L-lactic acid direct polycondensation products catalyzed by a non-metal-based compound. Journal of Applied Polymer Science. 2010;116(3):1597–602.
[22] Auras RA, Lim L-T, Selke SEM, Tsuji H. Poly (Lactic Acid): Synthesis, Structures, Properties, Processing, and Applications. Hoboken, NJ: John Wiley & Sons, 2010.
[23] Moon SI, Lee CW, Taniguchi I, Miyamoto M, Kimura Y. Melt/solid polycondensation of L-lactic acid: An alternative route to poly (L-lactic acid) with high molecular weight. Polymer. 2001;42(11):5059–62.
[24] Auras RA, Lim L-T, Selke SEM, Tsuji H. Poly (Lactic Acid): Synthesis, Structures, Properties, Processing, and Applications. Hoboken, NJ: John Wiley & Sons, 2011.
[25] Moon SI, Lee CW, Taniguchi I, Miyamoto M, Kimura Y. Melt/solid polycondensation of l-lactic acid: An alternative route to poly(l-lactic acid) with high molecular weight. Polymer. 2001;42(11):5059–62.
[26] Zhang W-x, Wang Y-z. Synthesis and properties of high molecular weight poly (lactic acid) and its resultant fibers. Chinese Journal of Polymer Science. 2008;26(04):425–32.
[27] Fukushima K, Furuhashi Y, Sogo K, Miura S, Kimura Y. Stereoblock poly(lactic acid): Synthesis via solid-state polycondensation of a stereocomplexed mixture of poly(L-lactic acid) and poly(D-lactic acid). Macromolecular Bioscience. 2005;5(1):21–29.
[28] Xu H, Luo M, Yu M, Teng C, Xie S. The Effect of Crystallization on the Solid State Polycondensation of Poly (L-lactic Acid). Journal of Macromolecular Science, Part B. 2006;45(4):681–87.
[29] Garlotta D. A literature review of poly (lactic acid). Journal of Polymers and the Environment. 2001;9 (2):63–84.
[30] Ajioka M, Enomoto K, Suzuki K, Yamaguchi A. The basic properties of poly (lactic acid) produced by the direct condensation polymerization of lactic acid. Journal of Environmental Polymer Degradation. 1995;3(4):225–34.
[31] Ajioka M, Suizu H, Higuchi C, Kashima T. Aliphatic polyesters and their copolymers synthesized through direct condensation polymerization. Polymer Degradation and Stability. 1998;59(1):137–43.
[32] Kim KW, Woo SI. Synthesis of high molecular weight poly (L-lactic acid) by direct polycondensation. Macromolecular Chemistry and Physics. 2002;203(15):2245–50.
[33] Cheng Y, Deng S, Chen P, Ruan R. Polylactic acid (PLA) synthesis and modifications: A review. Frontiers of Chemistry in China. 2009;4(3):259–64.
[34] Carothers WH, Dorough GL, Natta FJv. Studies of polymerization and ring formation. x. the reversible polymerization of six-membered cyclic esters. Journal of the American Chemical Society. 1932;54(2):761–72.
[35] Lowe CE. Preparation of high molecular weight polyhydroxyacetic ester. Google Patents, 1954.
[36] Monga S, Kaushik A, Gupta B. Optimization of controlled ring-opening polymerization of l-lactide to hydroxyl terminated polylactides using zinc acetate catalyst. Polymer Bulletin. 2017;74(12):4943–58.
[37] Kricheldorf HR. Syntheses and application of polylactides. Chemosphere. 2001;43(1):49–54.
[38] Kricheldorf HR, Kreiser-Saunders I. Polylactones, 19. Anionic polymerization of L-lactide in solution. Die Makromolekulare Chemie. 1990;191(5):1057–66.
[39] Yuan C, Xu X, Zhang Y, Ji S. Ring-opening polymerization of L-Lactide by an Anionic Cobalt (II) Aryloxide. Chinese Journal of Chemistry. 2012;30(7):1474–78.

[40] Haba O, Furuichi N, Akashika Y. Anionic ring-opening copolymerization of l-lactide with a five-membered cyclic carbonate having a glucopyranoside structure. Polymer Journal. 2009;41 (9):702–08.
[41] Albertsson AC, Palmgren R. Cationic Polymerization of 1,5-Dioxepan-2-one with Lewis Acids in Bulk and Solution. Journal of Macromolecular Science, Part A. 1996;33(6):747–58.
[42] Kricheldorf HR. Syntheses and application of polylactides. Chemosphere. 2001;43(1):49–54.
[43] Ragheb RT. Synthesis and Characterization of Polylactide-siloxane Block Copolymers as Magnetite Nanoparticle Dispersion Stabilizers. Blacksburg, VA: Virginia Polytechnic Institute and State University, 2005.
[44] Storey RF, Sherman JW. Kinetics and mechanism of the stannous octoate-catalyzed bulk polymerization of Îµ-caprolactone. Macromolecules. 2002;35(5):1504–12.
[45] Stridsberg K, Ryner M, Albertsson A-C. Dihydroxy-terminated poly(l-lactide) obtained by controlled ring-opening polymerization: Investigation of the polymerization mechanism. Macromolecules. 2000;33(8):2862–69.
[46] Stridsberg K, Albertsson A-C. Ring-opening polymerization of 1,5-dioxepan-2-one initiated by a cyclic tin–alkoxide initiator in different solvents. Journal of Polymer Science Part A: Polymer Chemistry. 1999;37(16):3407–17.
[47] Frediani M, Oberhauser W, Rosi L, Passaglia E, Bartoli M. Palladium-based catalysts-supported onto end-functionalized poly (lactide) for C-C double and triple bond hydrogenation reactions. New Advances in Hydrogenation Processes: Fundamentals and Applications. 2017;1. doi: 10.5772/65447.
[48] Singhvi MS, Zinjarde SS, Gokhale DV. Polylactic acid: Synthesis and biomedical applications. Journal of Applied Microbiology. 2019;127(6):1612–26.
[49] Hu Y, Daoud WA, Cheuk KKL, Lin CSK. Newly developed techniques on polycondensation, ring-opening polymerization and polymer modification: Focus on poly (lactic acid). Materials. 2016;9(3):133.
[50] Maharana T, Mohanty B, Negi YS. Melt-solid polycondensation of lactic acid and its biodegradability. Progress in Polymer Science. 2009;34(1):99–124.
[51] David DJ, Staley HB. Analytical Chemistry of the Polyurethanes. Hoboken, NJ: Wiley-Interscience Publications, 1969.
[52] Hood MA, Wang B, Sands JM, La Scala JJ, Beyer FL, Li CY. Morphology control of segmented polyurethanes by crystallization of hard and soft segments. Polymer. 2010;51(10):2191–98.
[53] Nakamae K, Nishino T, Asaoka S. Sudaryanto. Microphase separation and surface properties of segmented polyurethane; Effect of hard segment content. International Journal of Adhesion and Adhesives. 1996;16(4):233–39.
[54] Xu M, MacKnight WJ, Chen CHY, Thomas EL. Structure and morphology of segmented polyurethanes: 1. Influence of incompatibility on hard-segment sequence length. Polymer. 1983;24 (10):1327–32.
[55] Mondal S, Martin D. Hydrolytic degradation of segmented polyurethane copolymers for biomedical applications. Polymer Degradation and Stability. 2012;97(8):1553–61.
[56] Caracciolo PC, de Queiroz AAA, Higa OZ, Buffa F, Abraham GA. Segmented poly(esterurethane urea) s from novel urea-diol chain extenders: Synthesis, characterization and in vitro biological properties. Acta Biomaterialia. 2008;4(4):976–88.
[57] Chan-Chan LH, Solis-Correa R, Vargas-Coronado RF, Cervantes-Uc JM, Cauich-Rodriguez JV, Quintana P, et al. Degradation studies on segmented polyurethanes prepared with HMDI, PCL and different chain extenders. Acta Biomaterialia. 2010;6(6):2035–44.
[58] Reghunadhan A, Thomas S. Polyurethanes: Structure, properties, synthesis, characterization, and applications. In: Polyurethane Polymers. Elsevier, 2017, 1–16.

[59] Selukar B, Parwe S, Mohite K, Garnaik B. Synthesis and characterization of linear polylactic acid-based urethanes using Tin modified solid cloisite-30B catalyst. Advanced Materials Letters. 2012;3 (2):161–71.

[60] Li C, Liu J, Li J, Shen F, Huang Q, Xu H. Studies of 4,4-diphenylmethane diisocyanate (MDI)/1,4-butanediol (BDO) based TPUs by in situ and moving-window two-dimensional correlation infrared spectroscopy: Understanding of multiple DSC endotherms from intermolecular interactions and motions level. Polymer. 2012;53(23):5423–35.

[61] Leja K, Lewandowicz Gy. Polymer biodegradation and biodegradable polymers – a review. Polish Journal of Environmental Studies. 2010;19(2):255–66.

[62] Drioli E, Nakagaki M, Makino K, Arakawa M, Kondo T. Preparation and in vitro degradation properties of polylactide microcapsules. In: Membranes and Membrane Processes. Springer US, 1986, 371–77.

[63] You Y, Min B-M, Lee SJ, Lee TS, Park WH. In vitro degradation behavior of electrospun polyglycolide, polylactide, and poly(lactide-co-glycolide). Journal of Applied Polymer Science. 2005;95(2):193–200.

[64] Wei Z, Liu L, Zhang M, Yang F, Qi M. Preparation and in-vitro degradation of polylactide and poly (L-lactide-co-glycolide). Journal of Biomedical Engineering. 2008;25(1):122–26.

[65] Suuronen R, Pohjonen T, Hietanen J, Lindqvist C. A 5-year in vitro and in vivo study of the biodegradation of polylactide plates. Journal of Oral and Maxillofacial Surgery. 1998;56(5):604–14.

[66] Bonartsev AP, Myshkina VL, Nikolaeva DA, Furina EK, Makhina TA, Livshits VA, et al. Biosynthesis, biodegradation, and application of poly (3-hydroxybutyrate) and its copolymers-natural polyesters produced by diazotrophic bacteria. Communicating Current Research and Educational Topics and Trends in Applied Microbiology. 2007;1:295–307.

[67] Trinca RB, Felisberti MI. Segmented polyurethanes based on poly(l-lactide), poly(ethylene glycol) and poly(trimethylene carbonate): Physico-chemical properties and morphology. European Polymer Journal. 2015;62:77–86.

[68] Ghobadi E, Heuchel M, Kratz K, Lendlein A. Simulation of volumetric swelling of degradable poly [(rac-lactide)-co-glycolide] based polyesterurethanes containing different urethane-linkers. Journal of Applied Biomaterials & Functional Materials. 2011;10(3):293–301.

[69] Brzeska J, Morawska M, Sikorska W, Tercjak A, Kowalczuk M, Rutkowska M. Degradability of cross-linked polyurethanes based on synthetic polyhydroxybutyrate and modified with polylactide. Chemical Papers. 2017;71(11):2243–51.

[70] Kucharczyk P, Pavelkova A, Stloukal P, Sedlarik V. Degradation behaviour of PLA-based polyesterurethanes under abiotic and biotic environments. Polymer Degradation and Stability. 2016;129:222–30.

[71] Marcos-Fernandez A, Abraham GA, Valentin JL, San Roman J. Synthesis and characterization of biodegradable non-toxic poly (ester-urethane-urea) s based on poly (ε-caprolactone) and amino acid derivatives. Polymer. 2006;47(3):785–98.

[72] Lim D-I, Park H-S, Park J-H, Knowles JC, Gong M-S. Application of high-strength biodegradable polyurethanes containing different ratios of biobased isomannide and poly (ε-caprolactone) diol. Journal of Bioactive and Compatible Polymers. 2013;28(3):274–88.

[73] Wang W, Ping P, Chen X, Jing X. Shape memory effect of poly (L-lactide)-based polyurethanes with different hard segments. Polymer International. 2007;56(7):840–46.

[74] Ping P, Wang W, Zhang P, Chen X, Jing X. Shape-memory and biocompatibility properties of segmented polyurethanes based on poly(L-lactide). Frontiers of Chemistry in China. 2007;2 (4):331–36.

[75] Fan X, Tan BH, Li Z, Loh XJ. Control of PLA stereoisomers-based polyurethane elastomers as highly efficient shape memory materials. ACS Sustainable Chemistry & Engineering. 2017;5(1):1217–27.

[76] Zia F, Zia KM, Nazli Z-i-H, Tabasum S, Khosa MK, Zuber M. Preparation of hydroxyethyl cellulose/ halloysite nanotubes graft polylactic acid-based polyurethane bionanocomposites. International Journal of Biological Macromolecules. 2020;153:591–99.
[77] Gu S-Y, Jin S-P, Gao X-F, Mu J. Polylactide-based polyurethane shape memory nanocomposites (Fe_3O_4/PLAUs) with fast magnetic responsiveness. Smart Materials and Structures. 2016;25 (5):055036.
[78] Zia F, Zia KM, Tabasum S, Khosa MK, Zuber M. Preparation of hydroxyethyl cellulose/halloysite nanotubes graft polylactic acid-based polyurethane bionanocomposites. International Journal of Biological Macromolecules. 2020 Jun 15;153:591–599.
[79] Gu S-Y, Chang K, Jin S-P. A dual induced self expandable stent based on biodegradable shape memory polyurethane nanocomposites (PCLAU/Fe_3O_4) triggered around body temperature. Journal of Applied Polymer Science. 2018;135(3):45686.

J. Prabhakaran, Amit Kumar, Mohd. Mujtaba Ahmed,
Harveer Singh Pali

5 Prospects of biolubricants from nonedible Indian native feedstocks

5.1 Introduction

The automobile industry has been the focus of research efforts to improve its energy economy, dependability, and environmental friendliness [1]. Technological advancements such as lightweight components, exhaust gas treatment, greener fuels, and controlled combustion are being suggested to reduce the environmental impacts of machines and vehicles [2]. Lubrication is a process used to reduce wear between moving surfaces by injecting a lubricant material to assist in the dispersion of load and produce pressure between the surfaces. Lubrication aims to reduce wear and heat loss, protect against corrosion, minimize oxidation, act as an insulator in transformer applications, and act as a protective barrier against contaminants. Although wear and heat cannot be completely eliminated, their impact can be significantly reduced using lubricants, making them a useful tool. Friction can prevent heat and wear by reducing the coefficient of friction between surfaces. Lubricants, which come in various types like liquids, solids, and gases, are used to achieve this reduction, with liquid and solid/semisolid varieties being commonly used in daily situations [3]. Effective lubrication is crucial for the safe and reliable operation of a car, especially in engines and drive trains, to reduce wear and friction. Mineral oil, a by-product of crude oil distillation, was once used as a lubricant in automobiles. However, its limited supply limits its use and contributes to the destruction of aquatic and terrestrial ecosystems [4]. Burning mineral oil as a lubricant releases trace metal pollutants like calcium, magnesium, iron particles, and zinc, raising environmental concerns [5]. Mineral oil's future as a lubricant in automotive engines is uncertain, prompting the need for a viable replacement in the context of internal combustion engines, as predicted by Tung and McMillan [6]. The global search for innovation in substitute lubricants is fuelled by the depletion of crude oil stocks and rising oil costs, aiming to mitigate environmental damage caused by traditional lubricants and uncontrolled leaks [7]. Biolubricants are being promoted as a viable alternative to mineral oils due to their inherent technical properties and biodegradability [8]. Vegetable oils are ideal for boundary and hydrodynamic lubrication due to their long-chain fatty acids and polar groups, resulting in biolubricants with superior lubricity, higher flash point, lower volatility, and higher viscosity index (VI) compared to mineral oils [9]. Vegetable oils, due to their structural properties, including polar groups and extended chain fatty acids, are ideal for boundary and hydrodynamic lubrication [10]. Various renewable sources are being explored for the sustainable production of biolubricants [11]. The use of edible vegeta-

https://doi.org/10.1515/9783111145136-005

ble oil feedstocks is restricted due to concerns about their impact on human food supplies and environmental damage from traditional growing methods, and the potential for ecological imbalances when forest area is removed for planting [12]. Nonedible plant oils are being considered as a potential biolubricant option in the diesel engine industry [13]. Nonedible oils will be utilized as sustainable biolubricant feedstock due to their ability to be grown on wasteland without conflicting with food crops in limited agricultural regions. These nonedible plants are cost-effective, easily accessible, and offer similar or even more biolubricant and fuel characteristics to food crops [14]. Plant varieties like *Jatropha curcas*, polanga, karanja, neem, mahua, castor, jojoba, coriander, hingan, mongongo, and Kusum are known to produce nonedible vegetable oil seeds [15]. Nonedible oils are made from various raw materials such as tobacco, macaba, Abyssinian mustard, soapnut, milk bush, polanga, and Kusum oil [16]. The use of nonedible seed oils as biodiesel sources is increasingly emphasized, but there is a lack of research on the use of nonedible vegetable-based lubricants in the automobile industry, despite their widespread use as engine fuels.

The project aims to educate scientists, decision-makers, business people, and engineers on biolubricants, focusing on their potential to replace conventional lubricants in the automobile sector. It includes a comprehensive chapter on the subject, showcasing research papers from reputable publications, covering both current and previous works.

5.1.1 Origins and functions of lubrication

Lubrication is essential for equipment used in various sectors, involving applying lubricants between surfaces to reduce friction, prevent wear, control heat production, prevent corrosion and oxidation, and protect contact areas from impurities like dirt and dust. Lubricants also help control heat production, prevent corrosion and oxidation, and ensure efficient operation for extended periods. A thin barrier coating prevents wear and friction between surfaces after lubricant use, reducing wear and friction. This lubricant offers high thermal, oxidative, and corrosion resistance, with high flash and fire points, low pour point, high VI, boiling point, and freezing point, making it an excellent choice for various applications [17].

Researchers face the challenge of choosing the best lubricant for a specific system to enhance its reliability, durability, along with efficiency. The year 1883 marked a turning point in the history of lubrication with the use of oil for journal bearing lubrication [18]. Before crude oil's discovery, Egyptians used vegetable oils and animal-based fats as lubricants for equipment and vehicles, such as using olive oil-made calcium soaps on cartwheel axles [19]. Petroleum-based oils became dominant in the lubrication sector due to crude oil's exceptional lubricating properties and accessibility. This led to a booming industry and rapid consumption of crude oil. As demand increased, so did costs. To protect the environment and conserve fossil resources, interest in vegetable-based oils as lubricants increased. This led to numerous international businesses working on improving biolubricants, thereby broadening the scope of study in this area [20, 21].

5.1.2 Need for sustainable lubrication

Due to increased environmental awareness and legislation, the need for environmental-friendly lubrication solutions has grown. The depletion of scarce resources, environmental harm, and nonbiodegradability have all been associated with conventional petroleum-based lubricants. In contrast, biolubricants provide a greener option that is in line with international initiatives to cut carbon emissions and lessen environmental impact.

5.1.3 Biological oil explained

Bio-oils, derived from plants, vegetables, and animal fats, are composed of triacylglycerols, fatty acid molecules linked together. These oils have a thick consistency and viscosity due to their triglyceride structure. Bio-oils offer environmental safety, biodegradability, and renewable nature, with minimal environmental risks compared to mineral and synthetic oils. They are categorized into four types of fatty acids [8, 22]. Edible and nonedible oils are two types of vegetable oils, with edible oils being safe for human consumption and used in cooking. This chapter discusses the conversion of nonedible oils into biolubricants, which are often derived from plant waste, reducing concerns about food crop depletion for lubricant production. This approach offers a sustainable alternative to edible oils. Nonedible plants can thrive in both low and heavy rainfall areas, and their growth can be supported by cuttings and seeds. Farmers should consider incorporating nonedible crops into their cultivation efforts when arid sites are unsuitable for food production. This method can produce a significant amount of raw materials for biolubricants. Nonedible oils are renewable, biodegradable, and resistant to pests and illnesses. They have a high heat content and low sulfur content and can be extracted from various nonedible crops.

5.1.4 Why use biolubricants?

The depletion of crude oil sources and environmental concerns have led to a need for a successor with better qualities. The industrial revolution and modernization have increased energy needs, leading to increased reliance on petroleum fuels, resulting in higher fuel prices and pollution. To save energy and the environment, it is crucial to stop petroleum oil's rapid depletion. Additionally, there is a growing market for eco-friendly and biodegradable lubricants due to pollution caused by open lubrication systems. Bio-oils possess exceptional lubricating properties due to their high VI, enhanced viscosity, biodegradability, renewability, flash and fire points, nontoxic properties, low pour points, and strong metal adhesion. These properties make bio-oils a promising alternative to fossil fuels in lubricant production [23, 24]. Bio-oils are used to lubricate various mechanical systems, including motors, chainsaws, hydraulic systems, and steel industries. However, they often have reduced oxidation, thermal stability, and performance limitations at low temperatures. Proper additives can help mitigate these drawbacks, as they can be used as greases, metalworking fluids, and metal casting fluids [25, 26].

5.1.5 The conversion of biolubrication process

The following methods are frequently employed to transform unprocessed vegetable-based oils into biolubricants.

5.1.5.1 Pyrolysis

Pyrolysis is a thermochemical procedure that decomposes organic materials by subjecting them to heat treatment at extreme temperatures. This process results in the production of syngas, solid char, and liquid bio-oil [27]. This method has much promise for the treatment of municipal solid waste and organic waste since it has the potential to dramatically reduce the amount of trash, collect energy, and generate valuable secondary products. The malleability of pyrolysis makes it possible to modify the proportions of the products that are produced, providing a multi-faceted approach to the transformation of waste. It has garnered recognition for its capacity to handle environmental concerns and encourage innovation in environmentally responsible waste management solutions. Pyrolysis technology is a flexible and promising approach for converting materials via temperature and heating rates, changing low-energy-density materials into high-energy fuels. This is accomplished by heating the substance at a high temperature [28]. It takes an integrated approach to addressing difficulties associated with waste management and energy generation. The methods that have been developed are adapted to certain kinds of applications and waste, which ensure that the processing will be carried out effectively and that the final products will be clear.

Pyrolysis methods are versatile in that they may be used for a variety of feedstocks, and they can produce useful by-products such as biochar, bio-oil, and syngas from biomass, as well as resources from waste plastic and rubber. Figure 5.1 shows the schematic view of pyrolysis reactor. This technology helps to creating an economy that is more circular and efficient in its use of resources [29]. As research and development efforts continue, innovative reactor designs and optimized process parame-

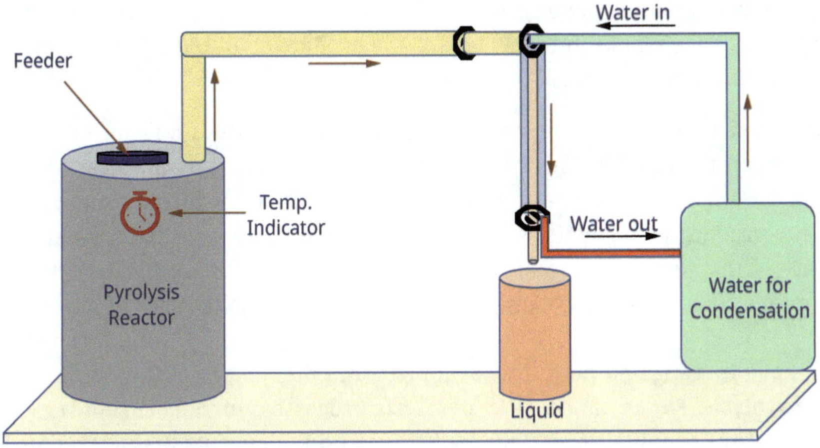

Fig. 5.1: Schematic view of pyrolysis reactor.

ters are anticipated to contribute to improved product yields and better efficiency. As a result, pyrolysis will become an essential component in plans for more environmental-friendly and sustainable waste management as well as energy generation. Equation (5.1) shows the components of pyrolysis products:

$$C_nH_mO_P(\text{feedstock}) \rightarrow (\text{heat}) \rightarrow \sum_{\text{liquid}} C_xH_yO_z + \sum_{\text{gas}} C_aH_bO_c + H_2O + C\,(\text{char}) \qquad (5.1)$$

The process of pyrolysis, which may be used for the conversion of a wide variety of waste materials, can be divided into four separate types: slow, traditional, quick, and catalytic. Depending on the values of factors such as vapor residence time and operating temperatures, each type displays a distinct set of features.

Slow pyrolysis: Pyrolysis at low temperatures includes vapor residence durations that are much longer, frequently lasting minutes or even longer. The processes of carbonization and torrefaction are subcategories that further divide this category. The operating temperatures and vapor residence periods of these processes are what set them apart from one another. These factors, in turn, have an effect on the composition and attributes of the end products.

Conventional pyrolysis: Pyrolysis according to conventional methods is a catchall term for any process that doesn't neatly fit into either the slow or the rapid categories of pyrolysis. Because it includes a wide range of vapor residence periods and working temperatures, it may produce a wide variety of finished goods.

Fast pyrolysis: Pyrolysis at a rapid pace takes place with vapor residence periods that are often measured in milliseconds or seconds rather than minutes or hours. Flash and ultra-rapid pyrolysis are both varieties of this form of pyrolysis, which is further subdivided. The faster heating and shorter vapor residence periods that occur during fast pyrolysis are what ultimately lead to the production of bio-oil and gases as the principal products of this kind of pyrolysis.

Catalytic pyrolysis: To carry out pyrolysis catalytically, a catalyst is required to be present throughout the process. This catalyst plays an important part in lowering the amount of oxygen that is present in the bio-oil that is produced, which ultimately results in the end product having better stability, a higher energy density, and less acidity. The capability of this technology to alter the chemical composition of bio-oil demonstrates the promise that it has for enhancing both the quality and the usage of bio-oil.

Researchers have investigated both the advantages and drawbacks of these different methods of pyrolysis. For instance, slow pyrolysis processes, such as carbonization and torrefaction, are suitable for certain feedstocks and the outputs that are wanted for the product. On the other hand, fast pyrolysis allows for the quick generation of bio-oil and gases, making it an attractive option for the production of energy as well as applications involving biofuels. Catalytic pyrolysis is an example of a novel way to refining product qualities and enhancing overall process efficiency. In this method, a catalyst is added to the reaction mixture.

The application of catalytic pyrolysis demonstrates its potential for increasing the characteristics of bio-oil via lower oxygen content and enhanced stability, as shown by Qureshi et al. [30]. These researchers demonstrated this promise by documenting their findings. These developments highlight the continuous attempts to perfect pyrolysis processes and widen their usefulness in waste management and sustainable energy generation strategies. Specifically, these efforts are being brought to light by the aforementioned achievements.

5.1.5.2 Transesterification

Triglycerides are converted into glycerol and fatty acid methyl esters through a chemical reaction with alcohol or methanol, with catalysts being acids or bases. The end products have improved low-temperature performance, lubricating characteristics, lower pour point, and increased oxidation stability. Controlling the alcohol amount can improve the production of these compounds [31, 32]. Figure 5.2 displays the flowchart of the transesterification process.

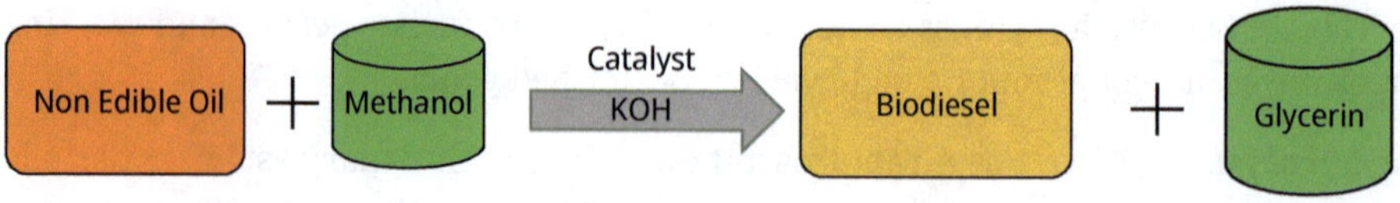

Fig. 5.2: Flowchart of transesterification process.

5.1.5.3 Hydrogenation

Hydrogenation is a chemical reaction using a nickel or platinum catalyst to inject hydrogen into vegetable oils. The properties of hydrogenated oils are influenced by double bond saturation, geometric isomerization, and positional isomerization. However, the addition of a nickel catalyst may increase oil safety. Figure 5.3 shows a flowchart of the hydrogenation process. There are two methods: partial hydrogenation and selective hydrogenation. This approach can be used for both partial and selective hydrogenation [33, 34].

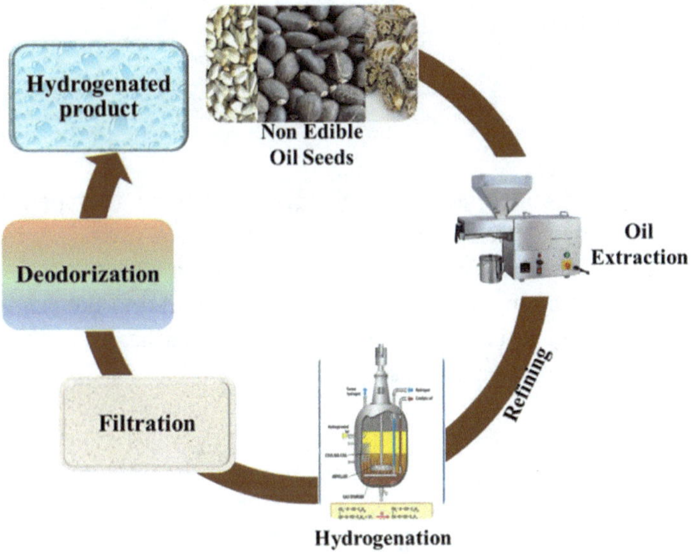

Fig. 5.3: Flowchart of the hydrogenation process.

5.1.5.4 Epoxidation

Epoxidation is a process that forms epoxides from alkenes and peroxy acids, resulting in epoxidized vegetable oils. These oils have improved antifriction properties but may not be durable at high temperatures. Figure 5.4 illustrates the experimental setup of epoxidation process. After epoxidation, the oil undergoes ring-opening and esterification procedures for further refinement [35, 36].

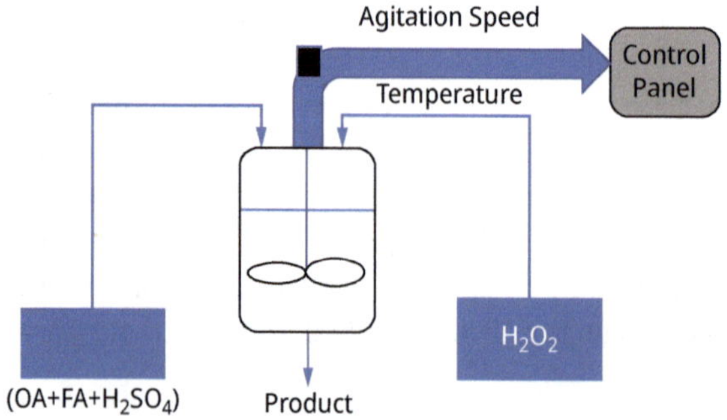

Fig. 5.4: Experimental layout of epoxidation process.

5.1.5.5 Direct additives mixing

Vegetable oils can be enhanced by adding additives in appropriate amounts, depending on their intended usage. These additives can be solids or liquids and are added in small amounts. Techniques like magnetic stirring or ultrasonication are used for a few hours to ensure proper integration. The final oil mixture, compared to the original base oil, shows better rheological and tribological properties. Careful selection and use of additives are crucial for optimal results.

5.1.6 Comparative analysis of physiochemical characteristics of nonedible oil

To have a better understanding of the many ways in which nonedible oils might be used in the manufacturing industry, it is necessary to conduct a comparative study of the physiochemical features of these oils. The term "nonedible oils" refers to oils that have been derived from plant sources but are not fit for human eating. These oils may be utilized for a variety of other applications, including the creation of biodiesel, industrial lubricants, cosmetics, and many more [37–39]. Table 5.1 displays the comparison of physiochemical properties of nonedible oil.

5.1.7 Types of additives used in lubricants

Pure bio-oils are in high demand, although they might have some restrictions in some areas. The inclusion of suitable additives is required to improve these qualities. Depending on the intended purpose, the amount of additives added to the basic oil might range from 0.1% to 30%. However, it is important to put their biodegradability

Tab. 5.1: Comparison of physiochemical properties of nonedible oil.

Property	Castor oil	Mahua oil	Jatropha oil	Polanga oil	Kusum oil
Botanical sources	*Ricinus communis*	*Madhuca indica*	*Jatropha curcas*	*Calophyllum inophyllum*	*Schleichera oleosa*
Color	Pale yellow	Pale yellow	Pale yellow	Pale yellow	Pale yellow
Density, 25 °C	0.962	0.901	0.91	0.887	0.923
Viscosity, 40 °C	8	62	26	34	42
Flash point (°C)	242	262	232	270	265
Pour point (°C)	−18	-9	−10	−5	−5
Saponification value (mg KoH/g)	182	191	193	192	194
Iodine value (g 12/100 g)	88	112	102	111	99
Fatty acid composition (%)	High in ricinoleic acid (92)	Palmitic [32] Oleic [43] Linoleic [22]	Palmitic [12] Stearic [14] Oleic [42]	Palmitic [22] Oleic [33] Linoleic [34]	Palmitic [18] Oleic [46] Linoleic [33]
Oxidative stability	Good	Moderate	Poor	Moderate	Good
Biolubricant potential	Yes	Yes	Limited	Limited	High

and nontoxicity first when selecting additives for bio-oils. It's crucial to avoid introducing hazardous compounds because doing so could undermine the principal benefit of employing biolubricants. There are several common categories of additives used.

5.1.7.1 Antioxidants

Lubricants contain antioxidants to slow down the oxidation process and increase their effectiveness and operating lives. Lubricating oils become more vulnerable to oxidation as operational speed and temperature increase, which subsequently causes additional friction and wear [40]. Numerous variables, including the quantity of oxygen and water in the oil, temperature, velocity, and the presence of wear debris, all affect the rate of oxidation [42]. Antioxidants are, therefore, added to the oil to slow down or prevent the oxidation process, increasing the oil's resistance to oxidation. Compounds like zinc dialkyldithiophosphate, zinc diamyldithiocarbamate, and amines are notable examples of efficient antioxidant additives [41].

5.1.7.2 Detergents and dispersants

These additives are essential for extending the life of the lubricant, increasing engine efficiency, and maintaining engine cleanliness in engine oils. Engine oils are continuously exposed to high temperatures and combustion by-products like sludge. These additives disperse the sludge particles, thereby preventing their accumulation, and reduce the generation of gases from incomplete combustion. Some additives perform both the duties of a detergent and a dispersant. Examples of typical compounds in this context include phosphates, sulfonates, phenates, calcium, and barium. It is crucial to remember that sulfonates and phenates should be avoided because they are dangerous and not environmental friendly [42].

5.1.7.3 Viscosity modifiers

Lubricants' viscosity is a crucial characteristic that directly affects how well they work. While excessively low viscosity may jeopardize the effective lubrication of sliding surfaces by enhancing their contact, excessively high viscosity may cause the temperature to rise and the buildup of foreign particles. As a result, a delicate balance between these extremes must be reached. Specialized additives called "viscosity modifiers" are applied to increase the viscosity of a lubricant. These modifiers typically appear in the form of fluids with unusually high or low viscosities. Notably, it has been found that ethylene-vinyl acetate copolymer can increase the viscosity of vegetable oils that have naturally low viscosities [43].

5.1.7.4 Viscosity index improvers

A lubricant's VI measures how its viscosity changes in response to temperature changes. A higher VI means that despite temperature changes, the lubricant's viscosity remains largely stable. Although bio-oils normally have high VI values, they can be improved even further by adding VI improvers such as ZnDDP and polymethacrylate. These additives increase the bio-low-temperature oil's and high-temperature viscosity characteristics, improving the bio-overall oil's lubricating effectiveness [44].

5.1.7.5 Rust and corrosion inhibitors

Sulfur, phosphorus, chlorine, iodine, oxygen, and water, among other elements, can cause corrosion, which can have a negative impact on mechanical parts. Therefore, it is crucial to get rid of these chemicals. Rust inhibitors like barium and calcium are used to protect ferrous metals against corrosion. To stop the corrosion of nonferrous metals, corrosion inhibitors such benzotriazole, zinc diethyl dithiophosphate, and zinc diethyl dithiocarbamate are also used. These inhibitors attach to the surface of the metal, forming a tough barrier [41].

5.1.7.6 Nanoparticles

Nanoparticles are a valuable addition for improving the performance of bio-based lubricants because of their exceptional abilities to reduce wear and friction. These nanoparticles create what is referred to as a "nanolubricant" when added to lubricants. A nanolubricant's efficacy is dependent on elements like the size, concentration, and shape of its nanoparticles. To avoid clustering, it is essential to make sure that the nanoparticles are carefully blended and evenly distributed throughout the oil. The accumulation of nanoparticles in the oil increases the wear on the three bodies of the abrasive. Notably, nanoparticles like TiO_2, ZnO, MoS_2, and CuO display outstanding effectiveness by reducing friction by providing a ball-bearing-like action between sliding surfaces. These nanoparticles work by interacting with the surrounding environment and the oil, which results in the development of a protective tribolayer on surfaces. By preventing direct material contact, this tribolayer acts as a barrier, lowering friction and wear [45].

5.1.7.7 Pour point depressants (PPDs)

The lowest temperature at which a liquid can still flow is known as its pour point. Bio-oils frequently perform poorly at low temperatures because they tend to freeze when the temperature falls below 0 °C. Pour point depressants (PPDs) can be added to base oils in very small amounts to help alleviate this problem. These PPDs work by preventing the growth of wax crystals, which makes it possible for oil to flow smoothly even in chilly temperatures. Alkyl naphthalene, poly-alpha-olefin, polymethyl acrylate, and other such substances are examples of such PPDs [46].

5.1.7.8 Friction modifiers

Friction modifiers are among the many sorts of additives included in bio-oils and are of utmost significance. These additives reduce the amount of friction that develops between a tribological pair's interacting surfaces. When applied, the additive particles act as tiny ball bearings by adhering to the surface. The overall amount of friction is decreased as a result of this change from sliding to rolling friction. Surprisingly, friction modifiers have substantial performance, especially when low temperatures and light loads are present.

5.1.7.9 Antiwear additives

These additives use chemisorption rather than straightforward adsorption, and are typically made of elements like phosphorus, sulfur, and chlorine [12]. They interact chemically with the surface and produce a remarkably tough layer that is resilient even at high temperatures.

5.1.7.10 Extreme pressure (EP) additives

These additives, which are specifically made to withstand high pressures and speeds, are intended to reduce friction and wear by depositing a thin layer of lower shear strength on the surface. Extremely high levels of these extreme pressure (EP) additives might cause corrosion, while extremely low levels cannot successfully reduce friction and wear to the necessary level. As a result, the precise concentration of these EP additives requires careful optimization. Examples include lead naphthalene, molybdenum disulfide, and related substances [41].

5.2 Current status of biolubrication market

Due to environmental concerns, strict regulations, and the need for sustainable substitutes for traditional lubricants made from fossil fuels, the biolubrication industry has seen consistent development. Biolubricants, commonly referred to as biolubricants or bio-based lubricants, are made from renewable resources such as plant and animal oils as well as other biodegradable substances. Compared to petroleum-based alternatives, these lubricants have a lower environmental effect, better biodegradability, and lower toxicity. The use of biolubricants has been accelerated by governmental legislation and policies that support sustainable practices, notably in the automotive and transportation industries. Due to growing consumer and manufacturer interest in lowering vehicle carbon footprints, the automotive sector is a prominent market driver for biolubricants. Table 5.2 illustrates the current global scenario of biolubricants.

However, there are some challenges, such as the greater price of biolubricants than traditional lubricants. Because bio-based lubricants must be produced using specific processing and refining methods, their general adoption in price-sensitive sectors may be constrained. Players in the market, both established and up-and-coming, are investing in R&D to produce novel biolubricant compositions with enhanced performance qualities. The biolubrication market is anticipated to keep growing through 2021 as businesses and consumers place more significance on greener options.

Tab. 5.2: Current market situation of biolubricants.

Aspect	Current situation and market trends for biolubricants
Market growth	The biolubricant market is experiencing growth due to the growing demand for sustainable and eco-friendly lubrication options.
Environmental focus	Climate change, resource depletion, and pollution are driving industries and consumers to adopt greener products due to growing concerns.
Regulatory support	Industries are implementing biolubricants to comply with strict regulations and requirements aimed at reducing carbon emissions.
Performance gains	The formulation of biolubricants has been enhanced due to improvements in oxidation stability, viscosity index, and temperature resistance.
Diverse applications	Biolubricants are increasingly utilized in various industries due to their increasing applicability and compatibility.
Consumer awareness	Rising environmental consciousness among consumers is driving the demand for sustainable products, thereby boosting the industry's expansion.
Research and innovation	Research is advancing the development of new biolubricant formulations to expand their potential applications and address their limitations.
Challenges	Challenges in biolubricant performance include compatibility with current machinery, cost issues, and ongoing research to enhance performance.
Future outlook	The market for biolubricants is anticipated to keep expanding as businesses work to lessen their impact on the environment along with consumers looking for eco-friendly products.

5.2.1 Forecast biolubricants demand growth by region between 2015 and 2030

Due to expanding economies, industrial development, and rising levels of automobile ownership, different areas are predicted to have different rates of increase in the global lubricants demand between 2015 and 2030. Because of the region's rapid industrialization and urbanization, especially in China and India, tremendous development is anticipated. Table 5.3 forecasts the biolubricants demand for 2015–2030. Owing to their developed markets, increased fuel economy, and developments in lubricant technology, North America and Europe are predicted to rise somewhat. Due to the expansion of the economy and the emergence of a growing middle class, Latin America and the Middle East are predicted to have modest growth. The rise of Africa's lubricants demand is anticipated to be modest owing to urbanization and infrastructural development.

Tab. 5.3: Biolubricants' forecast for 2015–2030.

Year	Demand (million barrels)	Forecast growth rate (%)
2015	100	–
2020	120	20.0
2025	150	25.0
2030	180	20.0

5.3 Challenges of biolubrication

Biolubricants have limitations due to their reduced effectiveness in cold weather and vulnerability to oxidation. They harden at below-freezing temperatures due to wax crystallization, and raw vegetable oils are susceptible to fast oxidation and not suitable for high-temperature settings. To address these issues, choose suitable chemicals and adjust the oil's pour point to improve low-temperature properties. Additionally, raise flash and fire points to enhance the lubricant's tolerance to high temperatures. The full utilization of nonedible resources faces challenges such as finding supplies from distant regions, dealing with subpar plants or seeds, limited availability, marketing issues, and technological constraints in harvesting and processing techniques [12].

5.4 Conclusion

Biolubricants have the potential to replace petroleum-based lubricants and contribute significantly to the lubrication sector. They can be extracted using simple techniques, making them readily available. Nonedible sources are preferred to prevent food crop depletion. Despite extensive research, bio-oils have not been widely used due to their inability to compete with synthetic and mineral oil-based lubricants. The wise selection of additives for specific applications is necessary to address these restrictions. With current research and development, nonedible bio-oils may become a significant lubricant source in the future.

References

[1] Ruggiero A, D'Amato R, Merola M, Valášek P, Müller M. On the tribological performance of vegetal lubricants: Experimental investigation on Jatropha Curcas L. oil. Procedia Engineering [Internet]. 2016;149:431–37. Available from: http://dx.doi.org/10.1016/j.proeng.2016.06.689

[2] Ong HC, Silitonga AS, Masjuki HH, Mahlia TMI, Chong WT, Boosroh MH. Production and comparative fuel properties of biodiesel from non-edible oils: Jatropha curcas, Sterculia foetida and Ceiba pentandra. Energy Conversion and Management. 2013;73:245–55.

[3] Panchal TM, Patel A, Chauhan DD, Thomas M, Patel JV. A methodological review on bio-lubricants from vegetable oil based resources. Renewable and Sustainable Energy Reviews. 2017;70:65–70.

[4] Ssempebwa JC, Carpenter DO. The generation, use and disposal of waste crankcase oil in developing countries: A case for Kampala district, Uganda. Journal of Hazardous Materials. 2009;161 (2–3):835–41.

[5] Miller AL, Stipe CB, Habjan MC, Ahlstrand GG. Role of lubrication oil in particulate emissions from a hydrogen-powered internal combustion engine. Environmental Science & Technology. 2007;41(19):6828–35.

[6] Tung SC, McMillan ML. Automotive tribology overview of current advances and challenges for the future. Tribology International. 2004;37(7):517–36.

[7] Ruggiero A, D'Amato R, Merola M, Valašek P, Müller M. Tribological characterization of vegetal lubricants: Comparative experimental investigation on Jatropha curcas L. oil, rapeseed methyl ester oil, hydrotreated rapeseed oil. Tribology International. 2017;109:529–40.

[8] Erhan SZ, Asadauskas S. Lubricant basestocks from vegetable oils. Industrial Crops and Products. 2000;11(2–3):277–82.

[9] Jayadas NH, Prabhakaran Nair K, Ajithkumar G. Tribological evaluation of coconut oil as an environment-friendly lubricant. Tribology International. 2007;40(2):350–54.

[10] Alves SM, Barros BS, Trajano MF, Ribeiro KSB, Moura E. Tribological behavior of vegetable oil-based lubricants with nanoparticles of oxides in boundary lubrication conditions. Tribology International. 2013;65:28–36.

[11] Lathi PS, Mattiasson B. Green approach for the preparation of biodegradable lubricant base stock from epoxidized vegetable oil. Applied Catalysis B: Environmental. 2007;69(3–4):207–12.

[12] Atabani AE, Silitonga AS, Ong HC, Mahlia TMI, Masjuki HH, Badruddin IA, et al. Non-edible vegetable oils: A critical evaluation of oil extraction, fatty acid compositions, biodiesel production, characteristics, engine performance and emissions production. Renewable and Sustainable Energy Reviews [Internet]. 2013;18:211–45. Available from: http://dx.doi.org/10.1016/j.rser.2012.10.013

[13] Biresaw G. Elastohydrodynamic properties of seed oils. JAOCS, Journal of the American Oil Chemists Society [Internet]. Jun [cited 2023 Aug 14] 2006;83(6):559–66. Available from: https://link.springer.com/article/10.1007/s11746-006-1240-9

[14] Yang L, Takase M, Zhang M, Zhao T, Wu X. Potential non-edible oil feedstock for biodiesel production in Africa: A survey. Renewable and Sustainable Energy Reviews. 2014;38:461–77.

[15] Singh SP, Singh D. Biodiesel production through the use of different sources and characterization of oils and their esters as the substitute of diesel: A review. Renewable and Sustainable Energy Reviews. 2010;14(1):200–16.

[16] Hasni K, Ilham Z, Dharma S, Varman M. Optimization of biodiesel production from Brucea javanica seeds oil as novel non-edible feedstock using response surface methodology. Energy Conversion and Management. 2017;149:392–400.

[17] Singh Y, Farooq A, Raza A, Mahmood MA, Jain S. Sustainability of a non-edible vegetable oil based bio-lubricant for automotive applications: A review. Process Safety and Environmental Protection. 2017;111:701–13.

[18] Recovery E, Of P, Engine W. CHAPTER 2.:2–4.

[19] Norrby T. Environmentally adapted lubricants – Where are the opportunities?. Industrial Lubrication and Tribology. 2003;55(6):268–74.

[20] Environmentally Friendly and Biobased Lubricants. Environ Friendly Biobased Lubr. 2016 Sep 19;

[21] Mobarak HM, Niza Mohamad E, Masjuki HH, Kalam MA, Al Mahmud KAH, Habibullah M, et al. The prospects of biolubricants as alternatives in automotive applications. Renewable and Sustainable Energy Reviews. 2014;33:34–43.

[22] De Almeida SCA, Belchior CR, Nascimento MVG, Ldsr V, Fleury G. Performance of a diesel generator fuelled with palm oil. Fuel. 2002;81(16):2097–102.

[23] Tamada IS, Lopes PRM, Montagnolli RN, Bidoia ED. Biodegradation and toxicological evaluation of lubricant oils. Brazilian Archives of Biology and Technology. 2012;55(6):951–56.

[24] Rani S, Joy ML, Nair KP. Evaluation of physiochemical and tribological properties of rice bran oil – Biodegradable and potential base stoke for industrial lubricants. Industrial Crops and Products. 2015;65:328–33.

[25] Luna FMT, Cavalcante JB, Silva FON, Cavalcante CL. Studies on biodegradability of bio-based lubricants. Tribology International. 2015;92:301–06.

[26] Salih N, Salimon J. A review on eco-friendly green biolubricants from renewable and sustainable plant oil sources. Biointerface Research in Applied Chemistry. 2021;11(5):13303–27.

[27] Wathakit K, Sukjit E, Kaewbuddee C, Maithomklang S, Klinkaew N, Liplap P, et al. Characterization and impact of waste plastic oil in a variable compression ratio diesel engine. Energies. 2021;14(8):1–18.

[28] Loganathan M, Madhavan VM, Arun Balasubramanian K, Thanigaivelan V, Vikneswaran M, Anbarasu A. Investigation on the effect of diethyl ether with hydrogen-enriched cashew nut shell (CNS) biodiesel in direct injection (DI) diesel engine. Fuel. [Internet] 2020;277(April):118165. Available from: https://doi.org/10.1016/j.fuel.2020.118165

[29] Chen WH, Peng J, Bi XT. A state-of-the-art review of biomass torrefaction, densification and applications. Renewable and Sustainable Energy Reviews [Internet]. 2015;44:847–66. Available from: http://dx.doi.org/10.1016/j.rser.2014.12.039

[30] Qureshi MS, Oasmaa A, Pihkola H, Deviatkin I, Tenhunen A, Mannila J, et al. Pyrolysis of plastic waste: Opportunities and challenges. Journal of Analytical and Applied Pyrolysis. 2020;152:104804.

[31] Leung DYC, Wu X, Leung MKH. A review on biodiesel production using catalyzed transesterification. Applied Energy. 2010;87(4):1083–95.

[32] Madankar CS, Pradhan S, Naik SN. Parametric study of reactive extraction of castor seed (Ricinus communis L.) for methyl ester production and its potential use as bio lubricant. Industrial Crops and Products. 2013;43(1):283–90.

[33] Fernández MB, Sánchez M JF, Tonetto GM, Damiani DE. Hydrogenation of sunflower oil over different palladium supported catalysts: Activity and selectivity. Chemical Engineering Journal. 2009;155(3):941–49.

[34] Shomchoam B, Yoosuk B. Eco-friendly lubricant by partial hydrogenation of palm oil over Pd/γ-Al2O3 catalyst. Industrial Crops and Products. 2014;62:395–99.

[35] Tan SG, Chow WS. Biobased epoxidized vegetable oils and its greener epoxy blends: A review. http://dx.doi.org/101080/036025592010512338 [Internet]. [cited 2023 Aug 14] 2010;49(15):1581–90. Available from: https://www.tandfonline.com/doi/abs/10.1080/03602559.2010.512338

[36] Abdullah BM, Salimon J. Epoxidation of vegetable oils and fatty acids: Catalysts, methods and advantages. Journal of Applied Sciences. 2010;10(15):1545–53.

[37] Negi P, Singh Y, Tiwari K. A review on the production and characterization methods of bio-based lubricants. Materials Today: Proceedings. 2021;46(20):10503–06.

[38] Hamnas A, Unnikrishnan G. Bio-lubricants from vegetable oils: Characterization, modifications, applications and challenges – Review. Renewable and Sustainable Energy Reviews. 2023;182:113413.

[39] Wang R, Zhou WW, Hanna MA, Zhang YP, Bhadury PS, Wang Y, et al. Biodiesel preparation, optimization, and fuel properties from non-edible feedstock, Datura stramonium L. Fuel. 2012;91(1):182–86.

[40] Wanasundara PKJPD, Shahidi F. Antioxidants: science, technology, and applications. Bailey's Industrial Oil and Fat Products [Internet]. 2005;[cited 2023 Aug 14]; Available from: https://onlinelibrary.wiley.com/doi/full/10.1002/047167849X.bio002

[41] Engineering Tribology – Google Books [Internet]. [cited 2023 Aug 14]. Available from: https://www.google.co.in/books/edition/Engineering_Tribology/_wVoTz1pDlwC?hl=en&gbpv=1&dq=G.+Stachowiak,+A.W.+Batchelor,+Engineering+tribology,+Butterworth-Heinemann.++(2013).&pg=PR1&printsec=frontcover

[42] Rizvi SQA. 9.1.2 Compressor and Refrigeration Oils Page. Compr Rev Lubr Chem Technol Sel Des [Internet]. 2009 Jan 5 [cited 2023 Aug 14]; Available from: https://www.scribd.com/document/419792908/A-Comprehensive-Review-of-Lubricant-Chemistry-Technology-Selection-and-Design

[43] Quinchia LA, Delgado MA, Valencia C, Franco JM, Gallegos C. Viscosity modification of different vegetable oils with EVA copolymer for lubricant applications. Industrial Crops and Products. 2010;32(3):607–12.

[44] Rudnick LR, ed. Lubricant Additives : Chemistry and Applications, 2nd edition. 2009 Apr 20 [cited 2023 Aug 14]; Available from: https://www.taylorfrancis.com/books/mono/10.1201/9781420059656/lubricant-additives-leslie-rudnick

[45] Mushtaq Z, Hanief M. Enhancing the tribological characteristics of Jatropha oil using graphene nanoflakes. Journal of Tribology. 2021;28(February):129–43.

[46] A Comprehensive Review of Lubricant Chemistry Technology Selection and Design | PDF | Exhaust Gas | Oil Refinery [Internet]. [cited 2023 Aug 14]. Available from: https://www.scribd.com/document/419792908/A-Comprehensive-Review-of-Lubricant-Chemistry-Technology-Selection-and-Design

Priyanka Singh, Nathi Ram Chauhan

6 Indian scenario of various advanced alternative fuels in internal combustion engines

6.1 Introduction

Many nations that import more crude oil are concerned about the fast rising demands for fossil fuels and petroleum products. There is a demand for alternative fuel due to the rapid expansion of industrialization, the depletion of fossil fuel reserves, air pollution brought on by diesel engine exhaust emissions, and global warming. Consequently, the development of renewable energy sources is required. Alternative fuels have recently gained in popularity due to their advantages for the environment and utilization of renewable resources. Increased industrialization, economic liberalization, fierce rivalry, and mergers and acquisitions have altered the global landscape and unified people from all over the world [1, 2]. "Energy is the essential need for economic development of any country," and "petroleum diesel" is India's greatest source of energy after coal, with roughly two-thirds of it imported from OPEC (oil and petroleum exporting countries). This has provided plenty of opportunities for emerging nations to keep up with the pace of industrialization (oil and petroleum exporting countries) [18]. The Indian economy is unstable as a result of its high reliance on imported fuel and the regular price variations in the petroleum market, which have also led to inflationary trends that are wreaking havoc on the common man [18].

India imports about 85% of its petroleum oil requirements [21]. To cut down on oil imports, the Indian government has already made 10% ethanol blends in gasoline mandatory. The 20% ethanol-mixed gasoline specification for use as vehicle fuel is being finalized by the Bureau of Indian Standards [21]. In this context, this analysis discusses the existing and projected state of the Indian transportation, petroleum, and biofuel industries, as well as international developments on using ethanol and alcohol-based fuels as a substitute for gasoline in spark-ignited automobiles [18]. Various traditional references for information were collated, examined, and presented. By 2023, the assessment estimates that the global gasoline demand will be around 48 billion liters, and India can manufacture an additional 30 billion liters of ethanol annually [18, 21].

The number of vehicles has increased exponentially during the past ten years. Between 2006 and 2015, the number of vehicles rose from 90 million to 209 million, or almost 2.3 times more. Around 74% of all vehicles are two-wheelers, followed by passenger automobiles and freight vehicles. In 2010, there were 15 million vehicles sold worldwide; by 2015, there were 20 million (Society of Indian Automobile Manufacturers, 2017). Commercial vehicle sales have remained at about 0.7 million units, and during

https://doi.org/10.1515/9783111145136-006

the same time, sales of passenger cars rose from 2.5 to 2.8 million units. According to the Government of India's Automotive Mission Plan 2026, domestic vehicle sales would at least triple to 62 million units annually in 2026 from their current level, growing at a 12% annual rate of compound growth (India Brand Equity Foundation, 2017) [21]. According to the forecasts, India will have a sizable need for fuels for use in a vehicle over the course of the 10 years or the following decade due to a rapid rise in car sales [1]. The corporate average fuel consumption standard was set by the Indian government, trying to cut down on emissions of greenhouse gases and improve the fuel efficiency of automobiles (Ministry of Power, Government of India, 2015). April 2017 saw the implementation of the first group of regulations, while April 2023 will see the implementation of the second group of restrictions. The average weight of the vehicle when weighted depending on how many automobiles were sold under each model for each automaker will be used to compute the corporate average fuel consumption [21].

Automobile makers must cut fuel usage by up to 47% from 2009 to 2010 levels in order to fulfill the 2023 targets. Depending on the typical car weight, the proportion varies. It will be necessary to reduce fuel usage from 2017 levels by about 17% in order to fulfill the targets for 2023. Many technologies are being used to increase the fuel efficiency of cars, and the main ones are hardware modifications to reduce friction in the engine, the usage of engine oils with minimal viscosity, techniques for gasoline injection directly, engines with high compression ratios, and improvements to the quality of fuel. The gasoline's octane rating may be raised by adding ethanol to it. When 10% ethanol is added to gasoline, an increase in research octane number of about three units is often anticipated. Owing to better combustion, it would be possible for the engine to function at greater compression ratios while simultaneously reducing the consumption of fuel [19, 21].

The prime minister's announcement of the National Hydrogen Mission on the country's 75th Independence Day made it apparent that India was committed to promoting the use of green fuels. Additionally, automakers are making efforts to enhance the usage of renewable energy sources and adhere to various fuel efficiency standards. Green fuel has emerged as a viable option for the automotive sector to lower its carbon footprint as newer alternatives gain traction. As required by BS-VI (Bharat stage VI), green fuel will be essential in reducing both particulate matter and NO_x (nitrogen oxides) emissions (Bharat Stage Emission Standards) [21]. According to government policy, the use of alternative fuel technology would help India get closer to eliminating its reliance on crude oil imports. In addition to biofuels, E20 flex fuel (ethanol mixed) may potentially be an alternative shortly. Flex fuel will be widely used, which will considerably aid in pollution control. The leading automakers in India are adapting to society's evolving needs and aiding in a seamless, sustainable transition. Biodiesel, bioalcohol (methanol, E1510, butanol), ethanol–gasoline blends, methanol–gasoline blends, chemically based energy storage electricity batteries and fuel cells, fuel obtained from waste, hydrogen, nonfossil natural gas and methane (compressed natural gas, CNG), vegetable oils of various types, butanol, various types of algae oils, propane (LPG), synthesis gas, and different sources of biomass are a few popular alternative fuels [19]. Following that, the investigation demonstrates that the HRR (heat release rate) value is lower for every loading condition when there is less biodiesel in the fuel blends. In contrast, B20 provides the lowest HRR value while the engine is operating under heavier loading conditions [54]. K40 (by volume 60% diesel and 40% kerosene) and B20 (by volume 80% diesel and 20% biodiesel) have both undergone dead state analysis. Except K30, where it is noted to be greatest at 20% load, all blends and pure diesel had a little decline in exhaust gas energy when the load was increased [55].

After environmental and economic scenarios, there are also COVID-19 effects on biodiesel that affect the economy of the world. Due to a substantial slowdown in economic activity since the emergence of COVID-19 in late 2019 in China and its quick spread throughout the world, crude oil prices have dropped significantly [53]. The price of crude oil has also decreased as a result of competition among significant oil producers. According to analysis, the spread of COVID-19 might cause a 122–219 million gallon drop in the biodiesel market in 2020. To put this in perspective, this is comparable to 7–13% of production of biodiesel in the United States in 2019 [53].

The government of India is promoting the utilization of low-carbon or carbon-free fuels in energy power systems to fulfill emission goals and to achieve security of energy because primary drawbacks of using internal combustion engines to burn fossil fuels are their impact on degradation of environment, such as global warming [1, 2].

The two main national objectives should be taken into consideration when developing in India:

a) Increasing India's energy security by reducing its reliance on crude oil imports
b) Improving the environment – Lessening the effect of the transportation sector on CO_2 and tailpipe emissions.

6.2 Various alternative fuels

In internal combustion engines, a lot of different types of fuels are used. These fuels often fall into one of two categories: oxygen-containing fuel and/or fuel-containing methyl esters. Several fuels that contain oxygen, including diethyl carbonate, methyl tertiary butyl ether, methanol, and ethanol, have been utilized as fuel additives.

6.2.1 Biodiesel

An unpopulated diesel fuel with a cleaner burn made using sustainable energy sources is called "biodiesel" (monoalkyl esters) [19]. Biodiesel functions in combustion-ignition engines just like petroleum diesel does. Ethyl or methyl esters of fatty acids are used to make biodiesel, which is made from fresh or used vegetable oils and animal fats (both edible oils and nonedible oils). Nonedible oils derived from species of plants like Calophyllum inophyllum (Nag champa), *Jatropha curcas* (Ratanjot), *Hevea brasiliensis* (rubber), *Pongamia pinnata* (Karanj), and so on are main product sources for biodiesel in India [1, 2, 5]. Oil seeds including soybean, palm, and sunflower are used to make bio-oils [1]. Although there is no petroleum in biodiesel, it might be mixed in any amount that can be used with petroleum diesel fuel to make a mix with a variety of ratios or used in its pure form. Considering that biodiesel characteristics are equivalent to those of diesel fuels and gasoline fuels, these may be used in engines with compression ignition and little or no modification of the engine is needed. It doesn't need a specific infrastructure because it may be deposited like diesel fuel made of petroleum. By using biodiesel in place of traditional diesel fuel, undesired particulates, carbon monoxide, and unburned hydrocarbons are significantly reduced. However, based on the testing and duty cycle procedures, nitrogen dioxide emissions are either slightly decreased or significantly increased. Utilizing biodiesel decreased the amount of solid carbon in the particulate matter particles due to the oxygen in biodiesel, enabling more complete burning of carbon dioxide. Since biodiesel contains nearly no sulfur, no aromatic compounds, and around 10% incorporated oxygen that helps it burn completely, it is regarded as a clean fuel. Even when combined with petroleum fuel, its higher cetane number improves ignition quality. Because fresh oilseeds that are not edible cannot be used to make biodiesel, the cost of production has increased. Leftover cooking oils and animal fats are widely accessible at lower prices because of the significant growth of food industry [4, 5].

6.2.2 Methanol

Methanol (CH_3OH) is a clear, colorless alcohol fuel that can be produced using various sources, including coal, natural gas, biomass, and cellulose, and is also very cost-effective [19]. Dimethyl ether (DME) can also be produced from methanol. The Indian government (NITI Aayog) intends to use methanol for automotive purposes. India will transition to a "methanol economy," according to the Union Minister for Transport, in order to cut emissions and import oil costs. As of now, mostly coal or natural gas is used to make methanol using synthesis gas (H_2 + CO). It has a low solubility in fat and oil and doesn't mix with water, alcohols, esters, or any alternative organic solvents. Methanol has high fuel-to-air stoichiometry characteristics. It has a higher latent heat value and a high proportion of oxygen content. Greater oxygen content enhances combustion and lowers soot and smoke production. M-85, which contains 85% methanol and 15% gasoline, is the most common alternative methanol fuel used today. The usage of neat methanol (M-100) is potentially possible in the future [19]. Methanol can also be converted into MTBE (methyl tert-butyl ether), an ether that is used with gasoline to increase octane and produce oxygenated fuel. Neither complex organic species nor sulfur is present in methanol [13, 17].

6.2.3 Ethanol

Ethanol is an inexpensive, nonpetroleum fuel. E-85 is the main ethanol alternative fuel, similar to methanol [10, 19]. Bioethanol is ethanol that is produced using cellulosic biomass materials as opposed to conventional feedstock (starch crops). Bioethanol, most often utilized as biofuel, consists of biomass that contains components made of sugar such as grapes, sugar beet, sugarcane, and sugary sorghum. Additionally, it includes biomass that contains starch substances, including cellulosic substances like agricultural and forestry waste, bagasse, wood waste, and any other sources of sustainability such as commercial garbage. However, sugarcane molasses is used in making 98% of ethanol in India [21].

A second-generation ethanol biorefinery is being built by Hindustan Petroleum Corporation Limited (HPCL) to produce 30 million liters of ethanol annually, utilizing agricultural waste as a feedstock [21]. Both feedstocks containing sugar and those containing starch can be utilized to produce ethanol through fermentation. In accordance with cost and output of producing ethanol, Cellulosic feedstocks and/or agricultural waste able to be employed in processes of gasification and fermentation. By gasifying and then synthesizing the petroleum wastes, which is high in carbon, like gasoline coke, ethanol able can be produced. A hydroxyl group, designated as –OH, is found attached to an atom of the carbon in the molecules of the chemical compounds known as ethanol (CH_3CH_2OH) [19].

6.2.4 Compressed natural gas (CNG)

The Indian government began a massive distribution effort to disseminate CNG throughout the nation, which allowed for the conversion of automobiles to CNG fuel. Vehicles that run on CNG were developed by manufacturers, and currently the majority of these vehicles are produced internally and adhere to all safety and pollution regulations. In India, there are currently 30 lakhs CNG automobiles. These CNG vehicles can replace 3% of the imported oil. There is a ton of potential left. It is compressed between 200 and 250 bars of pressure to allow storage within the fitted cylinders of automobiles because it has poor energy density as a fuel for vehicles, and as a result, it is known as CNG. In India, CNG is a fuel for the future, which is a substitute for usage in both diesel and gasoline engines, thanks to its recoverable resources, which total more than 690 billion cubic meters. Some of the appealing qualities of CNG as a car fuel are low exhaust pollutants, minimal upkeep, minimum noise, comfort of the driver, not susceptible to adulterate, and so on [19].

6.2.5 Liquefied natural gas (LNG)

Methane is liquefied at a low temperature of about −160 °C to produce liquefied natural gas (LNG), an alternative fuel that has low carbon content. Liquid storage increases energy density and allows for moderate level of pressure storage in compact tanks for fuel. The usage of LNG results in significantly lower emissions of pollutants, including SO_2, NO_2, and CO, making it the most environmental-friendly fuel. Compared to diesel, which has an emission factor of 74,100 kg/TJ, LNG has an emission factor of just 56,100 kg/TJ. If diesel is substituted with LNG, it is estimated that carbon emissions will drop by around 0.7 tons per 1,000 L. LNG can cut CO_2 emissions from diesel-powered vehicles by about 30%. A combination of hydrocarbons makes up natural gas, with methane gas (CH_4) making up 80–95% of its volume. It can be produced through gas wells or in tandem with the extraction of crude oil. It can be stored on moving vehicles as either liquid or in the form of compressed gas [19].

6.2.6 Hydrogen

Indirectly or when blended with diesel and CNG, hydrogen's unique properties and lack of carbon make it a suitable fuel for the clean and efficient functioning of reciprocating engines. For gasoline-based engines, hydrogen gas is a better fuel choice due to its high autoignition temperature (858 K) [14]. On the other hand, the fuel's low ignition energy made it susceptible to preignition and unplanned combustion, which frequently results in high combustion temperatures and the generation of nitrogen oxides (NO_x). However, internal combustion engine emissions could be lessened by using hydrogen. Hydrogen has a poor energy density per volume. As a result, hydrogen storage is difficult since it needs extreme conditions to be conveniently stored such as chemical reactions, low temperatures, and high pressures [50]. To get a sufficient driving distance of more than 300 miles with hydrogen as a fuel, it must be compressed at high pressures (350–700 bar) [16]. Cost and safety will be the key complication.

The fuel of the future for transportation with low-carbon emissions is hydrogen fuel. The ideal source of energy for decarbonizing transportation is hydrogen since it will improve energy security, lessen dependency on oil, cut greenhouse gas emissions, and clean the air [14, 16]. In addition to its cobenefits of reducing air pollution, affordability, and environmental-friendly transportation, hydrogen fuel might reduce greenhouse gases by 33–55% from their 2005 level by 2030 [17]. The largest per mass unit energy content with the existing fuels is found in hydrogen, which has 120.7 kJ/g. Hydrogen burns, leaving behind water as a by-product. One of the top research organizations focusing on hydrogen production from biomass connected to biological systems and additional forms of sustainable energy sources is Banaras Hindu University, along with the Indian Institute of Technology, Kharagpur, and Murugappa Chettiar Research Centre (MCRC), Chennai. New and Renewable Energy Ministry has provided research and development support to the MCRC, which has shown batch-scale hydrogen synthesis from distillery waste [14].

6.2.7 Synthesis gas (syngas)

Internal combustion (IC) engines can run on syngas as a fuel that is usable for both stationary generation of power and automobile applications [16]. It is produced primarily through the gasification and reforming of carbon-containing fuels like coal, heavy oil, biomass, and natural gas for stationary use, and chemically from liquid and gaseous fuels for automobile use. During these processes, complex and long-chain molecules are transformed into simpler ones like hydrogen, carbon monoxide, and methane for use in vehicles (nitrogen). The preferred composition often relies on the feedstock, feedgas, and production cost [16]. For instance, substantially sufficient purification is more expensive and time-consuming compared to the process of reaction reforming if using a process of biomass gasification to produce hydrogen as the main objective. Additionally, syngas can be created by fuel reformation using air and/or steam utilizing small equipment that might be installed within an automobile. The general goal of employing synthetic gas either as a primary fuel or a secondary fuel in internal combustion engines is to have the advantage of clean power conversion technology and renewable energy sources without compromising on the efficiency of engine, reduction in power, exhaust emissions, and particulate matter generation of engines powered by syngas. However, its benefits rely on the syngas' physicochemical characteristics (namely, how much CO and H_2 it contains), the combustion engine's type, and the method of fuel reforming utilized to produce syngas [16].

6.2.8 Dimethyl ether (DME)

Since it has a better cetane percentage, which is much higher than diesel and has minimal emissions of carbon monoxide, NO_x, and particulate matter, DME is a possible and viable energy substitute for petroleum or conventional fuels. By utilizing solid catalysts in catalytic processes, methanol can be dehydrated to form DME. To replace the usage of propane in home LPG, DME is produced by dehydrating methanol. DME blending with diesel can make it possible to replace the usage of diesel, giving us the chance to lessen our reliance on imported crude oil [16, 17].

6.2.9 Acetylene

Recent experimental studies on acetylene's usage in IC engines have been motivated by the substance's energy density and high flame speed. Compared to gasoline and diesel, acetylene burns more quickly and has a higher energy density and flame speed. Engines powered by acetylene may be closer to achieving the efficiency of a thermodynamically ideal engine cycle. However, acetylene has a lower octane rating compared to other alternative fuels for internal combustion engines [20]. As a result, acetylene consumption is only at its peak when knock first appears. The phenomenon of unwanted combustion and premature ignition is known as "knock," caused by fast flame speed, low ignition energy, low octane number, and broad flammability limitations. These are the primary issues with fueling internal combustion engines with acetylene. In SI engines, the cylinder is either directly or indirectly injected with gasoline and acetylene fuel; at the final stage of the compression stroke, a spark plug then ignites the mixture. It is either pumped directly into the cylinder in diesel engines or is introduced together with intake air. However, due to its extremely high temperature for self-ignition, the acetylene–air mixture does not ignite spontaneously. At the final stage of compression stroke, the mixture is injected with a small quantity of pilot fuel, a type of diesel fuel. Like a spark plug, the pilot fuel autoignites first to ignite the mixture of acetylene and air. The advantages of both spark ignition and compression ignition engines are thus integrated into dual fuel diesel engines [20].

6.2.10 Algae oil

Algae are another type of biofuel that has have a lot of potential. Algal of many types can be used to make algae oil. Algae can develop more quickly and can be produced under regulated conditions in a small space. A large number of people accept producing bio-oil from algae since it has a high oil content, is readily available, and can thrive in every atmospheric condition and environment including dams, rivers, ponds, and oceans [6, 22]. It may also flourish in flowing water such as canals used for agricultural water supply and urban sewers. Algae can conserve energy as carbohydrates, proteins, and lipids since they are effective photosynthetic creatures or organisms [6]. Additionally during photosynthesis, they convert carbon dioxide into energy very effectively, which contributes to their rapid growth. It has been determined that one acre can produce between 20,000 and 80,000 L of algal oil per year, the production in the amount of oil that can be produced through cultivating crops for a profit is more [6]. Algal oil's physiochemical characteristics make it ideal for internal combustion engines [6, 22].

6.2.11 Bio-butanol

To address various environmental and financial concerns, recently, bio-butanol fuel has served as a substitute fuel for internally combusted engines (gasoline or diesel engines) to conventional fuels [12]. This fuel is also known as "*n*-butanol" and it is an ecologically benign, renewable, and reasonably priced alternative fuel, similar to

Tab. 6.1: Combustion and physical properties of various alternative fuels.

S. no.	Properties	Gasoline [8, 12, 13, 14, 16, 20, 21, 25, 26, 38, 40, 44, 45]	Diesel [1, 4, 11, 12, 13, 14, 16, 20, 25, 26, 29, 38, 40, 43, 44, 45]	CNG [14, 16, 20, 25, 26, 38, 46]	Ethanol [13, 20, 21, 25, 26]	Hydrogen [14, 16, 20, 25, 26, 38, 43, 45, 46]	Acetylene [20, 37, 38, 42, 43, 44, 45, 46]	Methanol [13, 25, 26]	Propane (LPG) [25, 26, 40]	Biodiesel [1, 4, 7, 8, 11, 22, 26, 29]	Dimethyl ether [30]	Naphtha [41, 47]	Bio-butanol [8, 12]
1.	Formula	C_4–C_{12}	C_8–C_{20}	CH_4	C_2H_5OH	H_2	C_2H_2	CH_3OH	C_3H_8 majority and C_4H_{10} minority	Methyl esters of C_{12} to C_{22} fatty acids	CH_3OCH_3	C_nH_{2n} ($n = 5 \sim 8$)	C_4H_9 OH
2.	Density 1 atm, 20 °C (kg/m^3)	720–780	820–860	0.65	809.9	0.08	1.092	792	510–580	888	667	760–790	805
3.	Autoignition temperature (°C)	257	254	540	363	572	305	465	405–470	148	250	225	343
4.	Stoichiometric ratio (kg/kg)	14.7	14.5	17.2	9	34.3	13.2	6.45	15.70	12.4	9.0	–	11.2
5.	Motor octane number	95–97	–	105	89.7	130	45–50	88.6	96	–	–	58–62	85
6.	Flammability limits present in air (%vol.)	1.4–7.6	0.6–5.5	5.3–15	3–19	4–74.5	2.5–81	6.7–36	1.8–9.5	–	3.4–20	1.2–6.9	1.4–11.2

(continued)

Tab. 6.1 (continued)

S. no.	Properties	Gasoline [8, 12, 13, 14, 16, 20, 21, 25, 26, 38, 40, 44, 45]	Diesel [1, 4, 11, 12, 13, 14, 16, 20, 25, 26, 29, 38, 40, 43, 44, 45]	CNG [14, 16, 20, 25, 26, 38, 46]	Ethanol [13, 20, 21, 25, 26]	Hydrogen [14, 16, 20, 25, 26, 38, 43, 45, 46]	Acetylene [20, 37, 38, 42, 43, 44, 45, 46]	Methanol [13, 25, 26]	Propane (LPG) [25, 26, 40]	Biodiesel [1, 4, 7, 8, 11, 22, 26, 29]	Dimethyl ether [30]	Naphtha [41, 47]	Bio-butanol [8, 12]
7.	Adiabatic flame temperature (K)	2,300	2,200	2,320	2,082	2,254	2,500	1,949	1,967	~2,680	–	2,533	2,231
8.	Minimum quenching diameter (mm)	2.97	–	3.53	2.97	0.9	0.85	1.85	1.78	–	–	–	2.89
9.	Minimum ignition energy (MJ)	0.23	–	0.29	0.23	0.02	0.019	0.14	0.46	–	0.29	–	0.25
10.	Max. flame speed (m/s)	0.50	0.30	0.42	0.61	3.50	1.5	0.455	0.38	–	0.40	–	0.46
11.	Lower heating value (MJ/kg)	43.000	42.500	49.990	26.700	120.00	48.225	20.094	46.300	36.750	31.650	43.99	45.277
12.	Lower calorific value (MJ/kg)	44.5	42.5	47.1	26.96	120	49.9	19.66	46.6083	37.52768	28.88194	44.9383	33.1

13.	Higher calorific value (MJ/kg)	46.6	45.8	52.2	29.845	141.7	56.00	22.3	50.1532	40.16769	31.68012	48.0760	34.4
14.	Boiling point (is °C)	20–300	175–360	−162	78.3	−252.9	−84	64.8	−42	340–375	−24.81	26.7–148.9	118
15.	Flash point (°C)	−45	52–55	−184.4	17	–	−18.15	11	−104	100–170	−41	− 21.7	35
16.	Oxygen content (%)	0	0	0	34.73	0	0	50	0	0	34.8	0	21.5
17.	Cetane number	15–20	45	−10	8	–	–	5	<3	54	55–60	31–41	25

many other alternative fuels including natural gas, methanol, and ethanol. One of the finest fuels to replace traditional petroleum fuels (diesel or gasoline fuel) is considered to be bio-butanol. Bio-butanol has the capacity to lower automobile emissions and pollutants, which would benefit the atmosphere, ease strain on the supply of energy, and dramatically lower the reliance of the vehicle on resources that aren't renewable, and this is because bio-butanol performs better than traditional fuels (like diesel and gasoline) in several areas. Reviewing and evaluating the usage of bio-butanol in engines is important as the science behind this has made significant strides in recent years. Bio-butanol fuel can also be used by blending [12]. The combustion and physical properties of various alternative fuels are listed in Tab. 6.1.

6.3 Advantages of using alternative fuels in India in various sectors

- A rise in social well-being
- Environmental and health improvements for people
- Accessible and fair biodiesel policy
- Combating climate change and reducing emissions of greenhouse gases
- Fuel production and fuel supply within the country
- Technically viable oilseed crop farming, as well as handling and treating biodiesel
- Establishment of relationships and connections between all stakeholders involved
- Improvements in the environment, employment, and industry
- Reduced reliance and energy independence on imported fuel
- Technically viable oilseed crop cultivation and biodiesel production
- Development of vehicles that use these fuels
- Economic growth of the country

In Tab. 6.2, the Indian government has periodically established several policies governing the mixing of ethanol with gasoline in various ratios:

Tab. 6.2: Government policies on biodiesel blending in India [52].

S. no.	Year	Policies in India
1.	2000–2006	Biodiesel blending program
2.	2003	Government mandates 5% blending in nine states and four UTs
3.	2005	By 2017, the National Biodiesel Mission (NBM) aims to blend 20% biodiesel with conventional diesel

Tab. 6.2 (continued)

S. no.	Year	Policies in India
4.	2006	Government mandates 5% universal blending PAN India
5.	2006	A Biodiesel Purchase Policy was announced by MoP&NG (Ministry of Petroleum and Natural Gas) and went into effect
6.	2008	Government mandates 10% blending PAN India, vehicle made compliant with E10
7.	2008–2014	Biodiesel blending program
8.	2009	The national policy on biofuels set a 20% blending goal for 2017
9.	2012	10% blending is required to reach the 5% national average
10.	2013	To attain the national average of 5%, the government mandates blending of 10%
11.	2015	Government mandates informed usage of flex fuel for vehicles (E85 and ED95)
12.	2015	The government authorized the direct sale of biodiesel (B100) to bulk consumers like railroads and state road transportation corporations for blending with diesel
13.	2017	Average ethanol blending in India = 2.3% (target 10%)
14.	2018	Biofuel policy released, E20 by 2030
15.	2018	Galvanizing Organic Bio-Agro Resources (GOBAR) DHAN program
16.	2018	RUCO (Repurpose Used Cooking Oil) program
17.	2018	SATAT (Sustainable Alternative Towards Affordable Transportation)
18.	2019	From April 1, 2019, oil marketing companies must sell 10% EBP or any other percentage in accordance with BIS
19.	2019	Pradhan Mantri JI-VAN Yojana (JaivIndhan-Vatavaran Anukool Fasal Awashesh Nivaran Yojana)
20.	2019	Under the EBP Program (Ethanol-Blended Petrol), MoP&NG has released a "Long Term Ethanol Procurement Policy"
21.	2020	Atmanirbhar Bharat Abhiyan
22.	Presently	The country's biodiesel production is mostly based on imported palm stearin oil
23.	Presently	Initiatives for hydrogen from the MoP&NG

6.4 Conclusions

The following conclusions have been drawn from the present work:

- This research reviews the effects of different alternative fuel types' combustion and physical properties in internal combustion engines. This chapter concludes that a mixture of biofuels and alcohols is an excellent alternative fuel that per-

forms well in conventional engines and has better emission characteristics. They, therefore, seem to be preferable alternatives to diesel fuel.
- The Indian government has started by introducing the blending of gasoline with 10% ethanol because of the concerns regarding the consequences of pollution on the health of its citizens and the alarming rise in global warming.
- The Indian government is also working on a waste management system to find new alternative fuel by recycling wastes like waste cooking oil, waste plastic oil, and so on and also by recycling nanoadditives, which will be used in biofuel blending.
- In addition to reducing air pollution, this action has significantly decreased dependence on imported fuel, saving valuable foreign exchange. The Indian scenario for alternative fuels has been envisioned and analyzed in this study.
- Biodiesel is the best most sustainable option for petroleum-based gasoline and diesel engines. Because they produce more oil, third-generation biofuels can effectively reduce greenhouse gas emissions and replace fossil fuels.
- The manufacturing of biodiesel could become more affordable and sustainable with the help of fourth-generation biofuels and their fuel-making technology. The usage of biodiesel in vehicles can help the agricultural sector grow.
- The environment of our planet has been significantly harmed by the burning of fossil fuels. For this reason, alternative fuels come into the picture, but all physical and combustion properties vary for every alternative fuel in internal combustion engines. Therefore, it is crucial to analyze different fuel types to determine which ones are optimal in the short and long periods from the perspectives of economy and the environment.

References

[1] Kumar D, Prajapati A. Impact of biofuel in internal combustion engine. International Journal of Engineering Research and Technology. 2016;5(02).

[2] Kumar P, Sharma MP, Dwivedi G. Impact of biodiesel on combustion, performance and exhaust emissions of diesel engines. Journal of Integrated Science & Technology. 2014;2(2):57–63.

[3] Sanghamitra S, Deshmukh SS. Prominence of Bio-Fuels as an alternate fuel in CI engines. International Journal of Engineering and Advanced Technology. 2019;8:Issue–2S2.

[4] Gitay JS Prof, Selokar GR Dr. Feasibility and future prospects of biodiesel use in IC engines –A review. International Research Journal of Engineering and Technology. 2017;04(08).

[5] Channapattana SV Mr., Kulkarni RR Dr. Bio-diesel as a fuel in I.C. engines – A review. International Journal of Computer Science and Applications. 2009;2(1).

[6] Aravind S, Barik D, Ragupathi P, Vignesh G. Investigation on algae oil extraction from algae Spirogyra by Soxhlet extraction method. Materials Today: Proceedings. 2021;43(1):308–13.

[7] Sharma V, Duraisamy G, Cho HM, Arumugam K, AntoAlosius M. Production, combustion and emission impact of bio-mix methyl ester fuel on a stationary light duty diesel engine. Journal of Cleaner Production, JCLP. 2019;17172(233):147–59.

[8] Saraswat M, Chauhan NR. Comparative assessment of butanol and algae oil as alternate fuel for SI engines. Engineering Science and Technology, an International Journal 2215–0986. 2019;23(1):92–100.

[9] Venkata Sundar RK Mr., Kurbet SN Dr., Kuppast VV Dr. A review on performance of the IC engine using alternative fuels. Materials Today: Proceedings. 2018;5(1):1989–96.

[10] Yerrennagoudaru H, Manjunatha K, Vijaya Kumar KR, Dodda Basavanagouda M. Investigation and assessment of performance and emissions of an IC engine fuelled with diesel and different bio-fuels blended with Ethanol. Materials Today: Proceedings 5. 2018;5(2):Part 1 6238–46.

[11] Mahendran J, Saravanan K, Ragulnath D. Performance and emission characteristics of algae derived biodiesel processes. Materials Today: Proceedings. 2214–7853 Part 1. 2019;21:268–71.

[12] Zhen X, Wang Y, Liu D. Bio-butanol as a new generation of clean alternative fuel for SI (spark ignition) and CI (compression ignition) engines. Renewable Energy RENE 12488. 2019;147(P1):2494–521.

[13] Dabas N, Dubey V, Chhabra M, Dwivedi G. Performance Analysis of an IC Engine Using Methanol, Ethanol, and Its Blend with Gasoline and Diesel as a Fuel. Springer Nature Singapore Pte Ltd, 223–32, 2019.

[14] Dimitriou P. Taku Tsujimura, a review of hydrogen as a compression ignition engine fuel. International Journal of Hydrogen Energy. 2017;42:24470–86.

[15] Kumar N, Goel V, Chauhan S. Evaluation of the effects of engine parameters on performance and emissions of diesel engine operating with B40 biodiesel, Elsevier. Applied Energy. 2016;37(2):121–35.

[16] Paykani A, Chehrmonavari H, Tsolakis A, Alger T, Northrop WF, Reitz RD. Synthesis gas as a fuel for internal combustion engines in transportation. Progress in Energy and Combustion Science. 2022;90:100995.

[17] Tian Z, Wang Y, Zhen X, Liu Z. the effect of methanol production and application in internal combustion engines on emissions in the context of carbon neutrality: A review. Fuel. 2022;320:123902.

[18] Papreja AP Prof. (Dr.). Need for alternative fuel in India: A study. International Journal on Emerging Technologies Special Issue NCETST-2017. 2017;8(1):16–18.

[19] Mahesh Babu Talupula NV, Srinivasa Rao P, Kumar BSP, Praveen C. Alternative Fuels for Internal Combustion Engines: Overview of current research. SSRG International Journal of Mechanical Engineering (SSRG-IJME). 2017;4(4):20–29.

[20] Ilhak MI, Tangoz S, Akansu SO, Kahraman N. Alternative Fuels for Internal Combustion Engines, book-the Future of Internal Combustion Engines, chapter 3, 2019.

[21] Sakthivela P, Subramanian KA, Mathai R. Indian scenario of ethanol fuel and its utilization in automotive transportation sector. Resources, Conservation and Recycling. 2018;132:102–20.

[22] Monisha Miriam LR, Kings AJ, Edwin Raj R, Adhi Viswanathan M. Algal oil extraction-cum-biodiesel conversion in a novel batch reactor and its compatibility analysis in IC engine at various CRs. Fuel. 2021;293:120449.

[23] DOE, US Department of Energy. Alternative Fuels Data Center. see. http://www.afdc.energy.gov/fuels/fuel-comparison-chart.pdf. 2014.

[24] Ambade BU, Bhele SK, Nandgaye DB. experimental determination of flammability limit and quenching diameter for various biodiesel blends. International Research Journal of Engineering and Technology (IRJET). 2015;02(09).

[25] Bae C, Kim J. Alternative fuels for internal combustion engines. Proceedings of the Combustion Institute. 2017;36(3):3389–413.

[26] Holmborn J. Alternative fuels for internal combustion engines, The Swedish Energy Agency, see https://www.lth.se/, 2015.

[27] Janssen. The Impact of Different Biofuel Components in Diesel Blends on Engine Efficiency and Emission Performance, SAE Technical paper 2010-01-2119, 2119, 2010.

[28] Christensen. Properties and performance of levulinate esters as diesel blend components. Energy Fuels. 2011;25:5422–28.
[29] Satputaley SS, Zodpe DB, Deshpande NV. Performance, combustion and emission study on CI engine using microalgae oil and microalgae oil methyl esters. Journal of the Energy Institute. 2016;90(04).
[30] Chahardowli M, Farajzadeh R, Masalmeh SK, Mahani H, Bruining H. a novel enhanced oil recovery technology using di methyl ether/brine: Spontaneous imbibition in sandstone and carbonate rocks, society of petroleum engineers SPE, 181340–MS, pp. 1–22, 2016.
[31] Chena R, Wanga H-T, Wang H. Different oxygen levels of dimethyl ether combustion influence numerical simulation. Procedia Engineering. 2012;31:934–40.
[32] Li A, Yu L, Lu X, Huang Z, Zhu L. Experimental and modelling study on autoignition of a Biodiesel/n-Heptane mixture and related surrogate in a heated rapid compression machine. Energy Fuels. 2019;33(5):4552–63.
[33] Agarwal AK. Biofuels (alcohols and biodiesel) applications as fuels for internal combustion engines. Progress in Energy and Combustion Science. 2007;33:233–71.
[34] Zhu L, Cheung CS, Huang Z. Impact of chemical structure of individual fatty acid esters on combustion and emission characteristics of diesel engine. Energy. 2016;107:305–20.
[35] Li A, Zhu L, Mao Y, Zhai J, Han D, Lu X, Huang Z. Surrogate formulation methodology for biodiesel based on chemical deconstruction in consideration of molecular structure and engine combustion factors. Combustion Flame. 2019;199:152–67.
[36] Wang W, Gowdagiri S, Oehlschlaeger MA. The high-temperature autoignition of biodiesels and biodiesel components. Combustion and Flame. 2014;161:3014–21.
[37] Bisen KP, Mehare MR, Mudliar PD. Acetylene as alternative fuel in I.C engine. International Journal of Scientific & Engineering Research. 2020;11(6):1771–82.
[38] Nataraja M, Kiran Kumar M, Manjunath K, Madhukumar K. Acetylene as an alternate fuel in modified 4-stroke spark ignition engine. International Journal of Innovative Research in Science, Engineering and Technology. 2018;7(Special Issue 7):351–60.
[39] Bruscaa S, Lanzafameb R, Garranob AMC, Messina M. On the possibility to run an internal combustion engine on acetylene and alcohol. Energy Procedia. 2014;45:889–98.
[40] Dhakar R, Tripathi A, Raj J. use of LPG in internal combustion engines-a state of art review. Journal of Emerging Technologies and Innovative Research (JETIR). 2016;3(8):58–66.
[41] Birlenbacher Str., Material Safety Data Sheet Naphtha, TESORO, see https://www.collectioncare.org/, 2011.
[42] Material Safety Data Sheet Acetylene, AIRGAS INC, see https://www.airgas.com/, 2008.
[43] Sharma PK, Kuinkel H, Shrestha P, Poudel S. Use of acetylene as an alternative fuel in IC engine. Rentech Symposium Compendium. 2012;1:19–22.
[44] Datey UR, Bukhari AA, Khanzade AA, Pathan SM, Bhagat J. Acetylene as an alternate fuel in I.C. Engine. International Journal of Research in Engineering, Science and Management. 2020;3(3):239–41.
[45] Koli SR, Hanumantha Rao YV. Acetylene an potential alternative fuel for stationary diesel engine. International Journal of Recent Technology and Engineering (IJRTE). 2019;8(2):5013–16.
[46] Nehe A, Pargaonkar P, Mahadik P, Patil R, Katagihallimath S. Experimental analysis of acetylene gas as an alternative fuel for S.I. Engine. International Research Journal of Engineering and Technology (IRJET). 2019;06(06):403–08.
[47] Ali OM, Alomar OR, Ali OM, Alwan NT, Yaqoob SJ, Nayyar A. Sameh Askar and Mohamed Abouhawwash, Operating of gasoline engine using Naphtha and Octane Boosters from waste as fuel additives. Sustainability. 2021;13:13019.
[48] Park W. Naphtha as a fuel for internal combustion engines. International Journal of Automotive Technology. 2021;22(4):1119–33.

[49] Nwaiwu CF, Nwufo OC, Igbokwe JO, Ogueke NV, Emmanuel E. Anyanwu, Numerical study of the effect of a biodiesel on the cylinder liner of compression ignition engine. Proceedings of the ASME ES2017–3380, V001T02A007. 2017;13.

[50] Alternative fuels data center by U.S. Department of Energy, Energy efficiency and renewable energy, see https://afdc.energy.gov

[51] Nguyen VH, Duong MQ, Nguyen KT, Pham TV, Pham PX. An extensive analysis of biodiesel blend combustion characteristics under a wide-range of thermal conditions of a cooperative fuel research engine, Sustainability. 2020;12:7666.

[52] Policy/Act/Regulations for alternative fuels by Indian government, Ministry of Petroleum and Natural Gas, Government of India, see https://mopng.gov.in

[53] Taheripour F, Mintert J. Impact of COVID-19 on the Biofuels Industry and Implications for Corn and Soybean Markets. PURDUE University, see https://ag.purdue.edu/commercialag/home/resource/2020/04,2020

[54] Choudhary P, Sachar S, Khurana T, Jain U, Parvez Y, Soni M. Energy analysis of a single cylinder 4-stroke diesel engine using diesel and diesel-biodiesel blends. International Journal of Applied Engineering Research ISSN 0973–4562. 2018;13(12):10779–88.

[55] Srivastava S, Chaubey H, Parvez Y. Performance evaluation of CI engine using diesel, diesel-biodiesel blends and diesel-kerosene blends through exergy analysis. ICCEMME 2019- IOP Conference Series: Materials Science and Engineering. 2019;691(01):012066.

J. P. Yadav, Pratyush, Devendra Singh

7 Activated biochar and its applications in the field of agriculture and environment

7.1 Introduction

Coal, an impure form of carbon, recovered by partially burning carbonaceous substances, is millions of years old. Fossil fuels, exploited from the earth, are the outcomes of anaerobic decomposition under high pressure and temperature. Biochar, as an amender, improves the quality of mine soil and helps accelerate the growth of agricultural crops. Biomass is available inside the earth in abundance and in different forms. Burning the biomass wastes like crop residue, paddy straw, leaves, wood etc., pollutes the environment. To control the climatic changes, long-term planning to reduce CO_2 emissions is required. In modern techniques of biochar production, CO_2 and other greenhouse gases are re-burned or reduced to harmless elements "IUCN." During the photosynthesis process, the plant takes CO_2 from atmospheric air and facili-

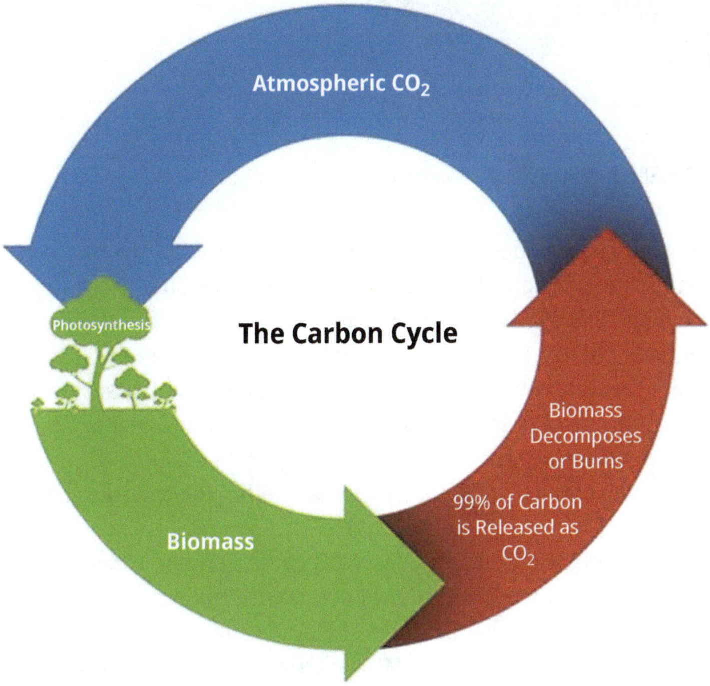

Fig. 7.1: Carbon returns.

https://doi.org/10.1515/9783111145136-007

tates it to grow, following the carbon cycle (Fig. 7.1). When the biomass is burned, most of the carbon returns to the atmosphere as CO_2.

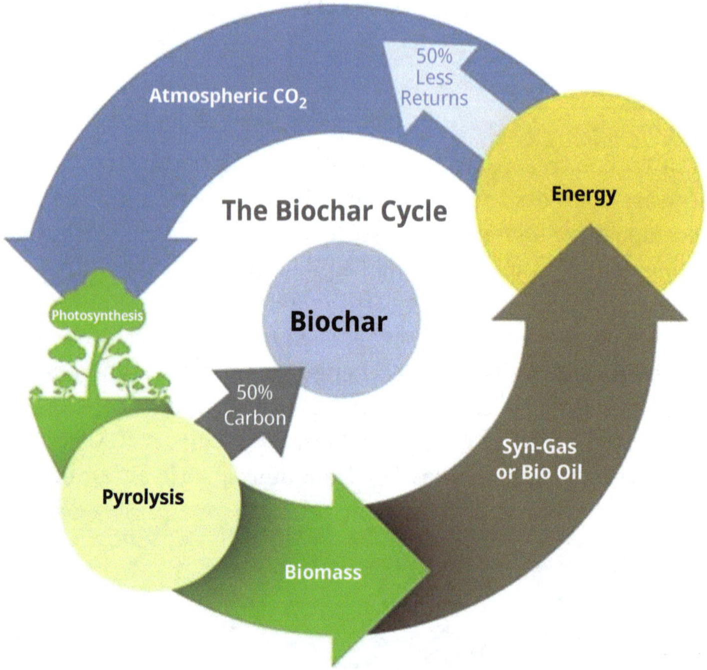

Fig. 7.2: Carbon sequestered to the air (problem) from atmosphere (solution).

Figure 7.2 shows that in biochar, produced as a consequence of photosynthesis, nearly 50% of the carbon remains there in solid biochar and the rest of the 50% is converted into producer gas and bio-oil "Fiat Panis" [1]. Biochar production through pyrolysis results in mitigation of climatic change and helps reduce the concentration of CO_2 in the atmosphere. Activated biochar, with larger surface area and porosity, facilitate the adsorption process. Hot gases are used to convert the source material into activated carbon. Different biochars, whose surface area lies in the range of 200–350 m^2/g, are not fertilizers, but act as nutrient transporters that provide shelter for microorganism growth.

The ancient and conventional methods of charcoal production are based on the formation of heap and cover it with wood and then burn it with limited access to air. This method, which is still used nowadays in some of the developing countries, creates a huge amount of smoke and releases about 50% of carbon from the original biomass as CO_2. Besides these traditional methods, the earth has its own created biochar known as "coal" that we find in the inner layers of the earth as a consequence of decaying of wood in absence of air at high pressure and temperature. The petroleum products, fossil fuels, and gases are the by-products of coal formation. For making biochar, two steel drums of different sizes are used as shown in Fig. 7.3; the smaller one is housed inside the bigger one and held on one or more bricks without touching the bottom.

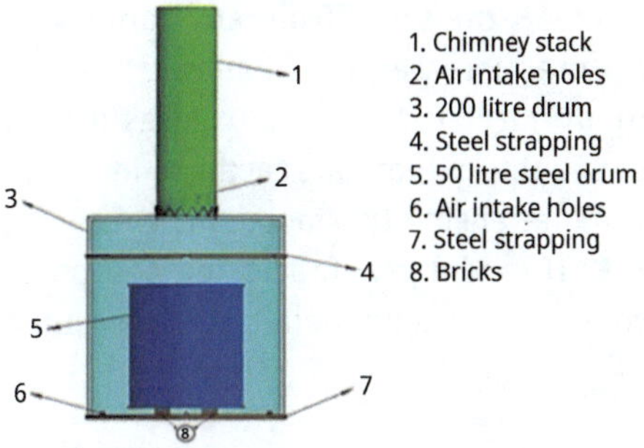

Fig. 7.3: Method of biochar production.

Using biomass to generate heat, in order to convert other biomass to biochar, seems like a very inefficient way of obtaining the biochar. Dry biomass is converted to biochar in the process that follows:

$$\text{Dry biomass} \rightarrow \text{char} + (CO,\ CO_2,\ H_2,\ H_2O\,(g),\ CH_4) + \text{tar} + \text{ash} \tag{7.1}$$

Kamali et al. [2] reviewed several literatures related to biochar amendment and reported that USA and China are global leaders in utilizing biochar for soils. Chen et al. [3] suggested that rice hull biochar stimulates the soil's microbial activity and increases the concentration of nitrogen and carbon. Yu et al. [4] reviewed the effect of biochar amendment in poor crops and reported that biochar amendment significantly improves the growth of the poor crop and enhances productivity. Xing et al. [5] reported that with the rice hull biochar amendment, the biomass of rice plants increases and MeHg concentration decreases in the pore water. Oladele [6] reported that rice-husk biochar reduces the soil's acidity and bulk density and increases water holding capacity and nutrient concentration. Yang et al. [7] observed that biochar amendment changes various components of leachate due to biochar properties like micropore volume, cation exchange capacity, total surface area, and labile matter concentration. De la Rosa et al. [8] reported that polycyclic biochar's of aromatic hydrocarbons are being affected by the pyrolysis temperatures and reactor type during their production. Thomazini et al. [9] reported that hardwood biochar reduces the formation of extractable nitrate concentrations and N_2O in different soils. Biochar amendment improves crop yield [10] and soil fertility [10, 11].

In the present work, experiments under atmospheric conditions were conducted to compare germination rate, growth rate, and water-holding capacity of biochar mixed soil for wheat crop and were observed after 11 days. The biochar effect on soil properties and crop was studied.

7.2 Materials and method

In the present work, an eco-friendly pyrolysis based experimental setup as shown in Fig. 7.4 was fabricated. The setup has a reactor that is a closed container of steel having high melting point and high pressure-bearing capability. The outer wall of the container is coated with Asian Royal paint that protects the setup from environmental effects and acts as an insulator. As the temperature increases, the biomass starts decaying to ensure carbonization. In this setup, an electric heater is used as a heat source and a good quality pressure gauge is used in the cylindrical reactor to measure the pressure during reactions. Pyrometer, a remote-sensing thermometer, is used to observe the temperature of the surface. The CAD model and fabricated setup are shown in Figs. 7.4 and 7.5, respectively. The experimental setup details are presented in Tab. 7.1.

Fig. 7.4: CAD model of experimental setup.

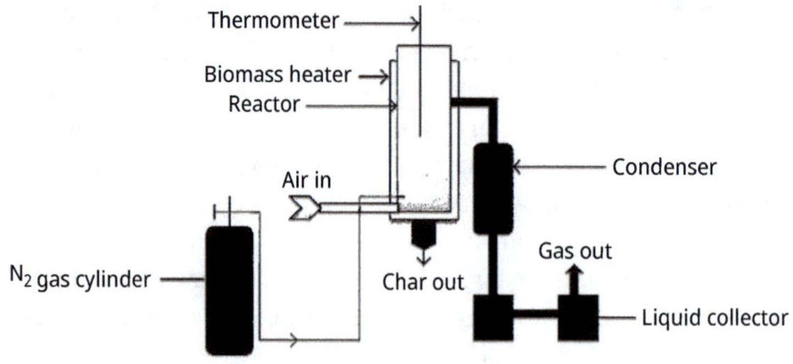

Fig. 7.5: Fabricated experimental setup for pyrolysis.

7.3 Result and discussion

The biomass used in the present work, for biochar production, is paddy straw made in the form of bricks by adding moisture to the paddy followed by drying and compression. Like paddy bricks, sugarcane bagasse bricks, wood chips, and pieces of coconut shells, as shown in Fig. 7.6(a)–(d), were also used for biochar production. The biochars, recovered from different biomass sources, were tested for their physicochemical properties from laboratories of NIT Rourkela (India). The test results are presented in Tab. 7.2. The variation of time, pressure, and temperature of biochar from different biomasses and their by-products (bio-oil and syngas) are recorded and

Tab. 7.1: Detailed specifications of the experimental setup.

Component	Material	Detailed specifications
Reactor	Stainless steel	Height = 25.40 cm Diameter = 24.63 cm Volume = 11,719.529 cm^3 Weight = 2110 g
Heat source	Nichrome	Height = 10.66 cm Diameter = 22.86 cm Capacity = 2 KW
Pressure gauge	Phosphor bronze	Range = 0–2 bar (0–30 psi) Dimensions = 6 × 4.1 × 2.3 cm Weight = 59 g
Pyrometer	Pyroelectric materials	Distance-to-spot ratio 30:1 Measuring temperature range ±1,050 °C
Ball valve	Cast iron	Diameter = 1.27 cm (inlet and outlet)
Exhaust pipe	Carbon steel	Diameter = 1.27 cm Pipe length (total) = 52.07 cm
Nut and bolts	Steel	M3-M10 as per requirement

Tab. 7.2: Report of physiochemical properties of biochar.

Biochar	pH	Electrical conductivity (μS)	Salinity (psu)	TDS	Na	Ca	K
				(ppm)	(mEq/g)	(mEq/g)	(mEq/g)
Wood	7.2	125	0.06	63	0.18	395.25	13.6
Paddy	8.8	523	0.27	222	1.14	2123	10.95
Coco peat	7.1	450	0.3	304	0.95	2080	21.1
Sugarcane bagasse	7.1	388	0.25	150	0.65	989	5.78

Source: Department of Biotechnology and Medical Engg, NIT, Rourkela, Odisha, India.

Tab. 7.3: Variation of time, pressure, and temperature of biochar formation.

Biomass	Time taken (min)	Highest temperature (°C)		Pressure
		Inner	Outer	
Paddy bricks	37	320	182	1 atm
Sugarcane bagasse	37	320	182	1 atm
Coconut shells	260	428	220	1 atm
Wooden pieces	260	428	220	1 atm

Tab. 7.4: Effects of biochar on soil components.

Samples	K (kg/ha)	N (kg/ha)	P (kg/ha)	pH	OC (%)	EC (ms/cm^2)
Sample 1: without biochar	175	222.02	11.7	6.9	0.34	0.18
Sample 2: with biochar	210	293.85	15.3	7.2	0.45	0.25
Standard data for wheat crop	101–250	333–522	20.1–40	6.5–8.5	0.51–0.8	<1

summarized in Tab. 7.3. A big difference in process time is noticed as different feedstocks have different carbon contents.

7.3.1 Analysis of soil properties before and after mixing biochar

Firstly, the dry soil sample is collected from the farm field and divided into two equal parts. One of the sample soils is mixed with the biochar (soil-to-char ratio of 7:3) and the other is kept unmixed. Then, both samples are tested in the soil testing laboratory under the Soil Health Card Scheme of the State and Central Government (India). The effect of biochar on soil components can be clearly seen from Tab. 7.4. The components of soil like K, N, P, pH, OC, and EC were variably increasing. Hence, this amended soil can be recommended for wheat crop as changed parameters are quite close to standard data for wheat crop as presented in Tab. 7.4.

7.3.2 Charging biochar

Biochar obtained from the biomasses used in the present work are activated for cultivation of wheat crop and water treatment by preparing a solution of 25% of calcium chloride ($CaCl_2$) by weight. Calcium chloride, a widely available nontoxic substance, is safer in handling even after adding water to it. To prepare 25% solution, three parts of water and one part of calcium chloride are mixed together. The required quantity of biochar to be charged (or activated) is mixed in the solution and left for 24 h. Now, the biochar is filtered and washed using distilled water. Then it is dried in high-

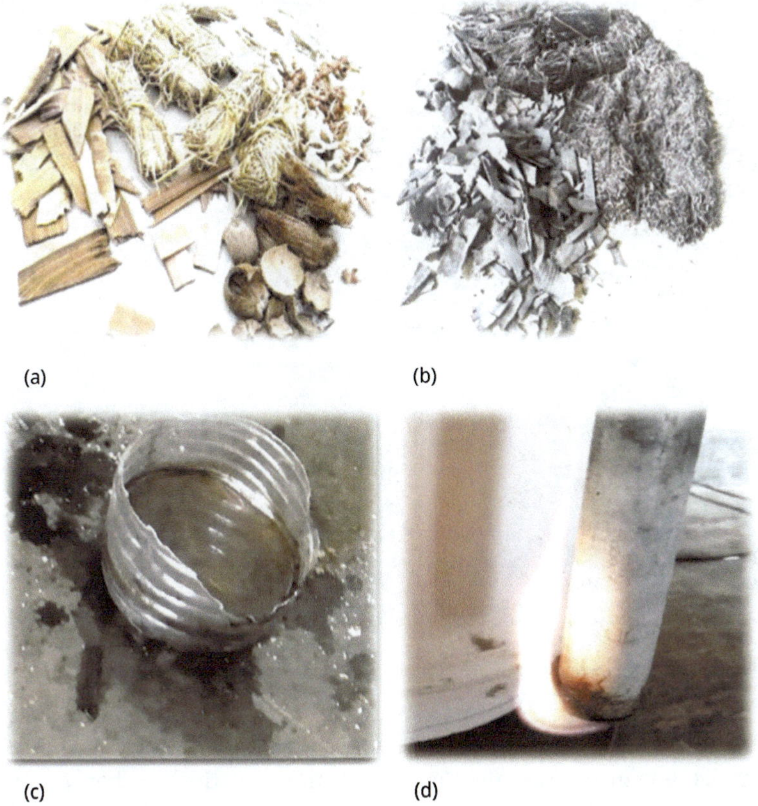

Fig. 7.6: (a) Different types of biomasses. (b) Biochar from different biomasses. (c) Bio-oil. (d) Burning flame of producer gas.

temperature environment using microwave oven. The activated biochar is prepared in accordance with the following equation:

$$\text{Biochar} + (CaCl_2 + H_2O) \rightarrow \text{activated biochar} \tag{7.2}$$

7.3.3 Applications of activated biochar

In the present work, six coconut shells were cut lesser than half from one side for filling the soil. Soil sample #1 (normal agricultural soil) was filled in three coconut shells. Soil sample #2 (soil mixed with activated biochar) was filled in the other three shells. At the base of all coconut shells, a small hole is made to drain extra water out from the shell. Wheat seeds, weighing 100 g, were dipped in water for 6 h and then filtered in a cotton cloth for another 6 h prior to sowing. Seeds were sown 1.5 cm deeper from the surface of the soil inside the coconut shells.

Fig. 7.7: Plant growth with/without biochar and effects when water intakes topped.

The experimental observations were made for 11 days. The germination started after a day in all six shells, but in some shells the germination was thicker than others though seeds were sown uniformly and at the same time. This is due to the soil amendment using biochar that improves the soil fertility. Hence, the growth of crops in biochar-mixed soil was found to be better than normal agricultural soil. The water retention ability of the soil was corroborated after the addition of activated biochar. The water intake was stopped after 7 days. It was observed that the crop without biochar was getting dried after 6 days when the water supply was stopped. As shown in Fig. 7.7, the crop starts getting dried after 7 days, and on day 11, it is completely dry. On the other hand, the crop containing biochar was not much affected by stopping the water supply. The growth of plants and effect of activated biochar are shown in Fig. 7.7. As shown in Fig. 7.7, the growth of the crops containing biochar, on daily basis, is better than the crop without biochar.

7.3.4 Water purification

The biochar produced in the present work was also used to purify the sewage water from a drainage canal. Initially, dark-colored sewage water collected from a sewage drain in a two-liter bottle was kept for sedimentation for 10 h. After ensuring the settling of mud, the sewage water was separated out in two bottles in equal quantity and labelled as sample #1 and sample #2. In sample #2, small pieces of activated biochar were filled and mixed by shaking the bottle. The samples were kept at a place for 12 h. The physical appearance of sample #2 was different in color and odor. This was attributed to the absorptive properties of the activated biochar; the toxic elements, oil and grease, were absorbed. Furthermore, the pH value of the water was also lowered. The parameters that were tested under particular protocol in the laboratory, along with their test results, are presented in Tab. 7.5.

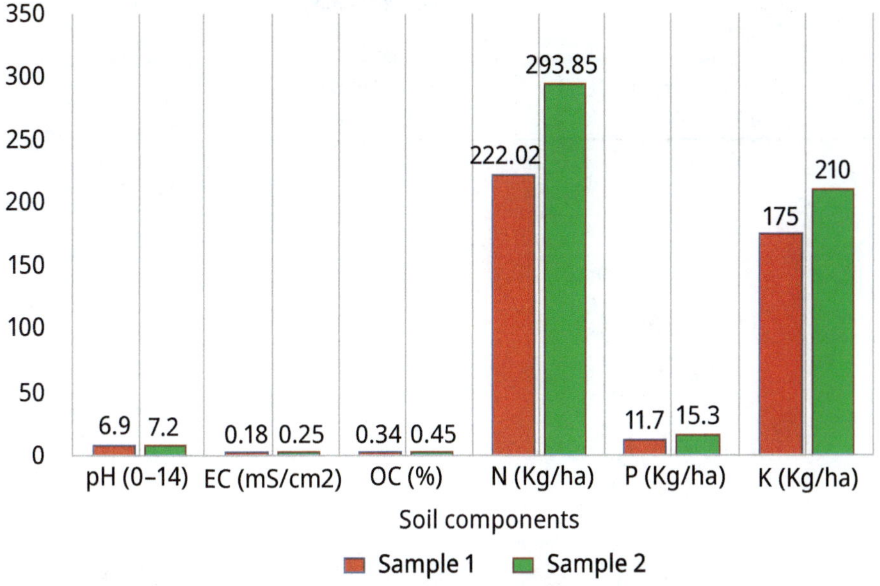

Fig. 7.8: Variation of soil parameters for sample 1 (without biochar) and sample 2 (with activated biochar).

Chemical parameters of normal sewage water (sample #1) and sewage water treated with activated biochar (sample #2), were tested; the values are presented in Tab. 7.6. The results are also plotted in Figs. 7.8 and 7.9 for comparative analysis. It can be revealed from the presented results that the application of activated biochar to treat sewage water significantly affects the values of pH, total suspended solids, oil, grease, chemical oxygen demand (COD), biochemical oxygen demand (BOD), and dissolved oxygen parameters. As shown in Fig. 7.8, an increase in the levels of N, P, and K is observed when activated biochar is added.

Tab. 7.5: Parametric analysis protocol.

S. no.	Parameters	Protocol
1	pH value	IS:3025 (pt-11)
2	Total suspended solids	IS:14345
3	Total dissolved solids	IS:3025 (pt-16)
4	Total hardness	IS:3025 (pt-21)
5	Oil and grease	IS:3025 (pt-44)
6	Chemical oxygen demand	SM (APHA)
7	Biological oxygen demand	IS:3025 (pt-44)
8	Iron	Spectrophotometer
9	Copper	AAS
10	Manganese	AAS
11	Dissolved oxygen	IS:3025(pt-39)
12	Organic matter	IS:3025(pt-44)

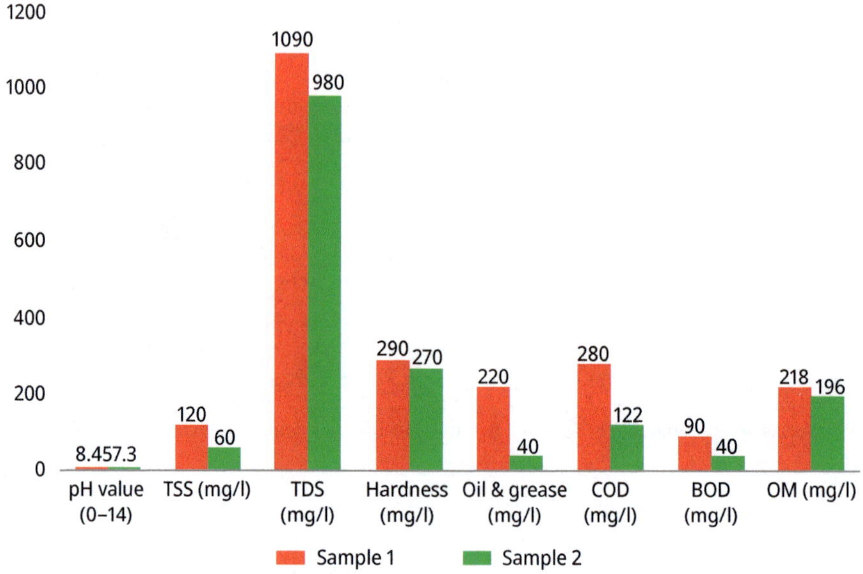

Fig. 7.9: Variation of sewage water parameters with and without activated biochar.

7.3.5 Effects of activated biochar on sewage water parameters

7.3.5.1 Effect of activated biochar on TDS value

The reduction in total dissolved solids (TDS) value, as plotted in Fig. 7.9, due to the presence of activated biochar, is attributed to the entrapment of solids on the biofilm that would have formed on the biochar throughout the adsorption process. Similar

Tab. 7.6: Sewage water parameters with and without activated biochar (sample 1: report no.: ORAS/NT/1206572; sample 2: report no.: ORAS/NT/1206573).

S. no.	Parameters	Sample 1	Sample 2
1	pH value	8.45	7.3
2	Total suspended solids (mg/L)	120	60
3	Total dissolved solids (mg/L)	1,090	980
4	Total hardness (mg/L)	290	270
5	Oil and grease (mg/L)	220	40
6	Chemical oxygen demand (mg/L)	280	122
7	Biochemical oxygen demand for 3 days at 27 °C (mg/L)	90	40
8	Iron (mg/L)	5.38	4.9
9	Copper (mg/L)	0.45	0.39
10	Manganese (mg/L)	1.45	1.25
11	Dissolved oxygen (mg/L)	3.8	2.09
12	Organic matter (mg/L)	218	196

observations were made by Manyuchi et al. [12] in which they have studied about the feasibility of biochar-treated sewage sludge.

7.3.5.2 Effect of activated biochar on COD and BOD levels

Activated biochar may play a vigorous role in environmental protection and can be efficiently used for sewage water treatment technologies. The sewage water treatment by activated biochar disinfects the water by trapping the oil, grease, and organic matter, and decreases the BOD and COD levels [13]. A decrement in COD and BOD levels, as shown in Fig. 7.9, was observed in the present work, when the sewage water is treated with developed biochar. This ensures a supply of cleaner water after being properly treated.

7.3.5.3 Effect of activated biochar on pH value

When the pH value of aqua solution is larger than the point of zero charge, negative charge exists on the surface of the adsorbent following the deprotonation of phenolic and carboxylic groups on its surface; then the competitiveness between the protons and the cation contaminants decreases [14, 15]. At lower levels of pH, basic functional groups such as amines get protonated and own a positive charge, which promotes the adsorption of anions [14, 15].

7.3.5.4 Effect of activated biochar in removal of inorganic pollutants

Inorganic pollutants in wastewater include heavy metals (Cr, Cu, Pb, Fe, and Zn as ions) and compounds like nitrate (NO_3), nitrite (NO_2), ammonium (NH_4), and hydrogen sulfide (H_2S), cause significant risk to public health and environment. Biochar, formed at lower pyrolysis temperature (<500 °C), exhibit properties suitable for the removal of inorganic compounds. The chemical composition and the morphological structure play an important role in the sorption nature of biochar [16]. In the present work, biochar prepared may have reduced the inorganic pollutants as a consequence of their physicochemical properties, which depends on the nature of the starting feedstock, the thermal conversion technique, and the preparation conditions [17].

7.4 Conclusions

In the present work, activated biochar, produced in the laboratory using the developed experimental setup for pyrolysis, is studied to analyze its effect in the fields of agriculture and environment. The following conclusions were drawn:

- Activated biochar, when mixed with agricultural soil, significantly affects the soil properties by improving the soil fertility, soil texture, alkalinity, pH, phosphorus, potassium, organic carbon percentage, and nitrogen.
- Addition of activated biochar improves the nutrient adsorption capacity of the crops. Consequently, the need for fertilizer is minimized, leading to increased growth and development of the crop.
- Biochar increases the crop production, yield, and water-holding (absorption) capacity. This in turn builds a resistance in the crop against the irregular irrigation cycle.
- Activated biochar may play a vital role in environmental protection and can be effectively used for sewage water treatment technologies. The sewage water treatment, using activated biochar, purifies the water by trapping the oil, grease and organic matter, and decreases the BOD and chemical oxygen demand COD levels. The by-products like syngas and bio-oil obtained during the conversion of biomass to biochar may be used as an energy source.

References

[1] Fiat Panis. Biochar – A Strategy to Adapt/Mitigate Climate Change in Climate Change Forum, accessed October 6, 2020.

[2] Kamali M, Jahaninafard D, Mostafaie A, Davarazar M, Gomes APD, Tarelho ML, Aminabhavi TM. scientometric analysis and scientific trends on biochar application as soil amendment. Chemical Engineering Journal. 2020; 395:125128.

[3] Chen S, Qi G, Ma G, Zhao X. Biochar amendment controlled bacterial wilt through changing soil chemical properties and microbial community. Microbiological Research. 2020;231:1263–73.
[4] Yu H, Zou W, Chen J, Chen H, Yu Z, Huang J, Gao B. Biochar amendment improves crop production in problem soils: A review. Journal of Environmental Management. 2019;232:8–21.
[5] Xing Y, Wang J, Shaheen SM, Feng X, Chen Z, Zhang H, Rinklebe J. Mitigation of mercury accumulation in rice using rice hull-derived biochar as soil amendment: A field investigation. Journal of Hazardous Material. 2019;388:121747.
[6] Oladele SO. Changes in physicochemical properties and quality index of an Alfisol after three years of rice husk biochar amendment in rainfed rice Maize cropping sequence. Geoderma. 2019;353:359–71.
[7] Yang XY, Chang KH, Kim YJ, Zhang J, Yoo G. Effects of different biochar amendments on carbon loss and leachate characterization from an agricultural soil. Chemosphere. 2019;226:625–35.
[8] De la Rosa JM, Martin AMS, Campos P, Miller AZ. Effect of pyrolysis conditions on the total contents of polycyclic aromatic hydrocarbons in biochars produced from organic residues: Assessment of their hazard potential. Science of the Total Environment. 2019;667:578–85.
[9] Thomazini A, Spokas K, Hall K, Ippolito J, Lentz R, Novak J. GHG impacts of biochar: Predictability for the same biochar. Agriculture, Ecosystems & Environment. 2015;207:183–91.
[10] Das SK, Ghosh GK, Avasthe R. Valorizing biomass to engineered biochar and its impact on soil, plant, water, and microbial dynamics: A review in Biomass. Conversion and Biorefinery. accessed, 2020.
[11] Yadav K, Tyagi M, Kumari S, Jagadevan S. Influence of process parameters on optimization of biochar fuel characteristics derived from rice husk: A promising alternative solid fuel. BioEnergy Research. 2019;12:1052–65.
[12] Manyuchi MM, Mbohwa C, Muzenda E. Feasibility of using sewage sludge bio char in treating municipal sewage. In: Proceedings of the International Conference on Industrial Engineering and Operations Management. Washington DC, USA, accessed, October 15, 2020.
[13] Mercado LFP, Lalander C, Berger C, Dalahmeh SS. Potential of biochar filters for onsite wastewater treatment: effects of biochar type, physical properties and operating conditions. Water. 2018;10:18–35.
[14] Kumar PS, Ramalingam S, Sathish Kumar K. Removal of methylene blue dye from aqueous solution by activated carbon prepared from cashew nut shell as a new low-cost adsorbent. Korean Journal of Chemical Engineering. 2011;28:149–55.
[15] Schwarzenbach RP, Gschwend PM, Imboden DM. Sorption I: General introduction and sorption processes involving organic matter. In: Schwarzenbach RP, Gschwend PM, Imboden DM, eds. Environmental Organic Chemistry. John Wiley &L Sons, Inc, 2002, 275–330.
[16] Abdelhafez AA, Li J. Removal of Pb (II) from aqueous solution by using biochar derived from sugarcane bagasse and orange peels. Journal of Taiwan Institute of Chemical Engineers. 2016;61:367–75.
[17] Enaime G, Baçaoui A, Yaacoubi A, Lübken M. Biochar for wastewater treatment conversion technologies and applications. Applied Science. 2020;10:3492.

Ajay Kumar Sharma, Devendra Singh, Shubham Mishra

8 An experimental analysis on double-slope single-basin solar still using phase change materials

8.1 Introduction

For life to exist and for living things to live on the Earth, water is essential and indispensable. Only 2.6% of the total amount of water is regarded as freshwater. This meager supply of freshwater makes it possible for life and vegetation to exist on the Earth. Globally, the consumption of freshwater is rising quickly, creating problems in the centuries to come [1]. Use of polluted water is a major contributor to waterborne illnesses. Due to water contaminants brought on by manufacturing activities and industrial operations, the amount of clean water on the Earth is continuously decreasing [2]. Due to the increase in global population, it is predicted that there will be a water deficit of around 60% by 2025 [3]. While water on the Earth contains a salinity of up to 10,000 ppm, the World Health Organization [4] states that the acceptable limit for salt in water is 500 ppm and for special instances up to 1,000 ppm. The total dissolved salt content of seawater typically ranges between 35,000 and 45,000 ppm. One environmentally friendly method of creating drinking water is through solar stills (SSs), which work by having water evaporate from a basin and condense on the glass above. Adding various materials to the basin of SSs is one approach to increase efficiency.

The SS has a low distillate production despite being relatively simple [5, 6]. Therefore, in order to boost the productivity of solar distillation units, researchers have looked into the different factors (climatic, operational, and designable aspects) that determine how well they function [7, 8]. Due to this, we have obtained several investigations on various distillation layouts, including stepped SS layout [9, 10], vertical SS [11, 12], tubular SS [13–15], SS with nanofluids [16–20], PV/T active SS [21, 22], hemispherical SS [23], finned SS [24–26], multistage SS [27], spherical SS [28], pyramid SS [29], and SS with phase change materials (PCMs) [30, 31].

El-Sebaii et al. [32] stated the noticeable improvement in daily distillate yield for SS combined with PCM from 0.64 to 2.38 kg/m^2-day as compared to a system without PCM. The primary benefits of PCM include its affordability, ease of accessibility, optimum temperature range for phase transitions, chemical stability over the long period and significant latent heat of conversion, and enhanced heat transfer and daily productivity.

To the best of our knowledge, no research has been done on the impact of quantity of PCMs (active and passive methods) on solar distillation using two different PCMs (paraffin wax and aluminum oxide) kept in a copper cylinder. A copper container is used to store PCM, and a limited amount of literature regarding solar distillation is also

https://doi.org/10.1515/9783111145136-008

available. We therefore examined the yields with PCMs and kept the PCMs in copper cylinders. Cylinders that are 0.1-cm thick are constructed in this investigation.

8.2 Experimental setup

The double-slope single-basin SS (DSSS) was designed and fabricated at the Institute of Engineering and Technology, Lucknow, as shown in Fig. 8.1. Galvanized iron (GI) sheet is used to make the SS, which has a bottom that is coated black to collect rich solar energy. The SS designed with DSSS was constructed using GI sheet and window glass cover 4-mm thick. The groundwater of Lucknow city is the brackish water used as the inflow water.

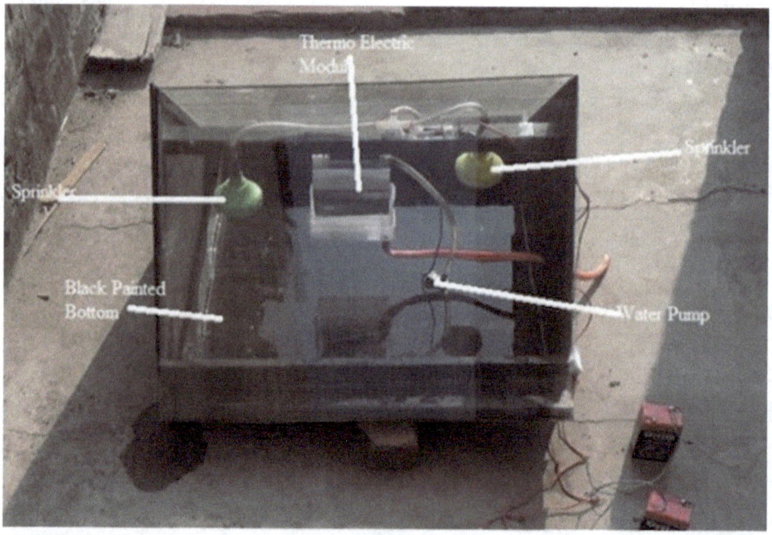

Fig. 8.1: Schematic of solar still.

The outside cover's dimensions are 37.5 cm × 32.31 cm × 76 cm, while the basin's dimensions are 76 cm × 75 cm × 10.2 cm. the region that lies between the basin and the glass-sheeted top covering of such hut top. GI sheet makes up the exterior portion. Two 4-mm-thick glasses are placed on top of the basin in such a way as to slope to the Sun's rays as they strike it from whatever direction that the still is placed. The temperature of the basin's bottom, as well as the glass surface and water layers equipped with insulator to stop loss of heat, was measured using thermocouples. The V-shaped spool that is built into the lower edge of the glass on both edges of the still is where the condensed water is collected. The condensate produced was gathered via flexible hose and kept in a measurement container. The PCMs (paraffin wax and aluminum oxide as shown in Figs. 8.2 and 8.3) are held in copper cylinders as shown in Fig. 8.4.

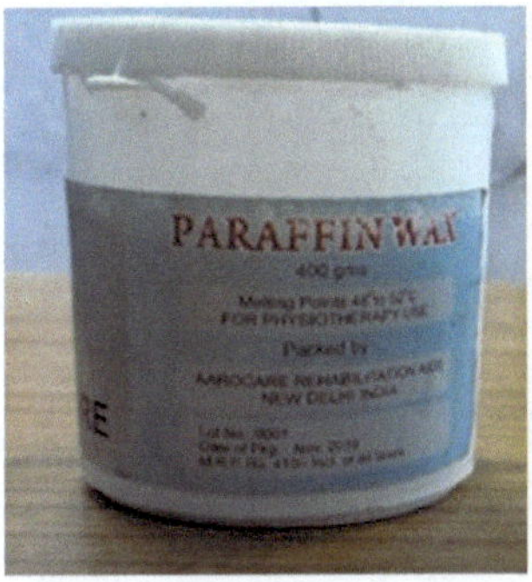

Fig. 8.2: Paraffin wax.

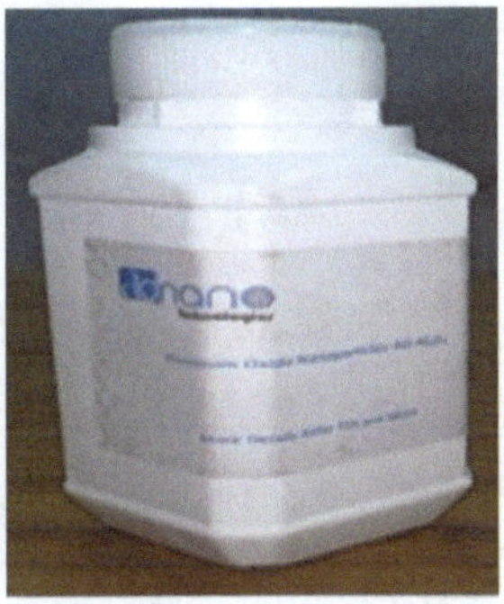

Fig. 8.3: Aluminum oxide.

Fig. 8.4: Copper cylinders.

The thermocouples can be embedded in the basin side divider through an opening, allowing for the measurement of the temperature of the inner glass surface, basin water, and vapor. At various locations, four thermocouples are placed in the basin. To measure the condensate temperature, two thermocouples are inserted into either side of the tank. To avoid the heat and vapor disaster, the space is sealed with protective material. Water tube from a piezometer is inserted through this gap to continuously provide crude water to the basin from the storage tank through a check valve that manages the flow, maintaining a constant quantity of water in the basin. In order to ensure that no heat is lost through water flow, a flexible rubber tube connecting hose is fitted between the water's inlet and output terminals.

Two water sprinklers are employed to increase the rate of evaporation as shown in Fig. 8.1. The water is being circulated by a USB submersible pump with dimensions of 4 cm × 3 cm × 2.5 cm, a flow rate of 4–5 L/h, and a voltage of 12 V with a current of 1.5 A.

Refrigeration can be accomplished using thermoelectric modules (Peltier) as shown in Fig. 8.5. The Peltier module, model number CP-14125840S, with dimensions of 40 mm × 40 mm, 450 P-N junction, 12 V, 60 W, and 5 A power, was employed. It was bought from Layered Technology. A solar power electric energy conversion system is employed to power such units to increase evaporation eventually and raise water temperature in order to increase production by receiving more radiation.

Due to the fact that smaller water droplets absorb less latent heat and evaporate more quickly, sprinkler arrangements have been made to obtain an atomized form of water. The thermoelectric module was utilized to create a cooling effect to raise the condensation rate, which in turn improved the water yield as shown in Fig. 8.1. Table 8.1 contains a list of the specifications for the parts.

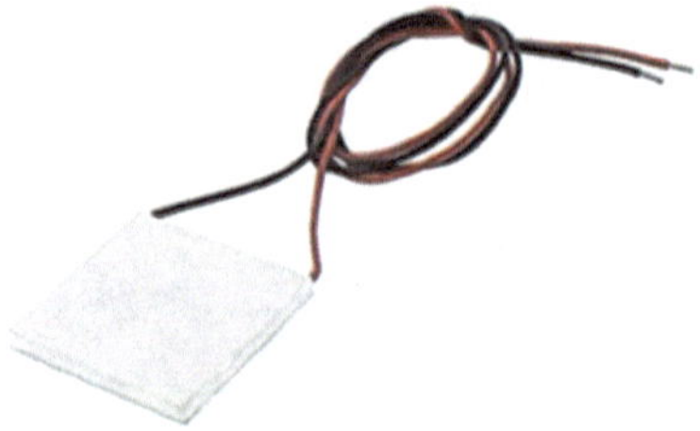

Fig. 8.5: Paraffin wax.

Tab. 8.1: List of the specifications for the parts.

Part	Model/suppliers	Specifications
Peltier module	CP-14125840S Layered Technology	Area: 40 × 40 mm^2 Current: 5A 450 P-N junction Volt: 12 V
Solar panel	LS1215 LuxmiSolar™	Size: 360 × 220 × 20 mm^3
USB submerge pump	Local market	Volume: 4 × 3 × 2.5 cm^3 Current: 1.5 A Flow rate: 4–5 L/h, Volt-12 V

8.3 Measurement tools

The accuracy, range, and uncertainty of the measurement tools are listed in Tab. 8.2.

Tab. 8.2: Measurement tools.

Tools	Range	Accuracy	Uncertainty
pH meter	0–14 pH	0.01 pH	0.006 pH
Thermocouple	0–100^{0}C	±0.1 °C	0.06 °C
TDS meter	0–1,000 ppm	10 ppm	0.1 ppm
Measuring jar	0–1,000 ml	±10 mL	0.1 mL

8.4 Heat transfer and energy balance equations

Energy transmission between the glass cover, PCM, copper cylinder, basin liners, and basin water is often what causes heat transfer in SS filled with PCM within copper cylinders. When the SS is charging and discharging, part of the solar energy is lost via the glass cover [33–38]. The main modes of heat dissipation through glass covers are convective, evaporative, and radiative, Dunkle [39] correlations can be used to compute the convective ($h_{c,w\text{-}g}$), radiative ($h_{r,w\text{-}g}$), and evaporative ($h_{e,w\text{-}g}$) heat transfer coefficients. When considering about the HT-EBEs, various assumptions have been made. These are listed below:

- The side losses are disregarded, and the system is thought to be vapor tight and fully insulated.
- In comparison to PCM and basin water, the heat capacities of insulating material, glass covers, copper cylinders, and basin liners are insignificant.
- The primary mechanism for heat exchange within the PCM and copper cylinder is conduction.
- The amount of heat that is transferred from the basin liners to the copper container and also from the copper container to the glass covers is not taken into account.
- As it is easier, the PCM does not take into account heat gradient.
- For all measurements, the average temperature has been considered.

8.4.1 Heat loss due to radiation

According to Dunkle and Chen et al. [39, 40], eq. (8.1) can be used to determine the rate of heat transmission by radiation between the surface of glass covers and mass of water:

$$Q_{r,w-g} = h_{r,w-g} \times \left(T_w - T_{gi}\right) \tag{8.1}$$

where

$$h_{r,w-g} = \varepsilon_e \sigma \left[\left(T_w + 273\right)^2 + \left(T_{gi} + 273\right)^2 \right] \left(T_w + T_{gi} + 546\right) \tag{8.2}$$

and

$$\varepsilon_e = \left(\frac{1}{\varepsilon_w} + \frac{1}{\varepsilon_g} + 1 \right) \tag{8.3}$$

8.4.2 Heat loss due to convection

Equation (8.4) allows for the calculation of the rate of heat transmission by convection within the SS:

$$Q_{c,w-g} = h_{c,w-g} \times \left(T_w - T_{gi}\right) \tag{8.4}$$

where

$$h_{c,w-g} = 0.884\left[\left(T_w - T_{gi}\right) + \frac{\{P_w - P_{gi}\}\{T_w + 273\}}{268,900}\right]^{0.333} \tag{8.5}$$

and

$$P_w = \exp^{\left(25.317 - \frac{5,144}{T_W + 273}\right)} \tag{8.6}$$

$$P_{gi} = \exp^{\left(25.317 - \frac{5,144}{T_{gi} + 273}\right)} \tag{8.7}$$

8.4.3 Heat loss due to evaporation

Equation (8.8) allows for the calculation of the rate of heat transmission by evaporation within the SS:

$$Q_{e,w-g} = h_{e,w-g} \times \left(T_w - T_{gi}\right) \tag{8.8}$$

where

$$h_{e,w-g} = 13.273 \times 10^{-3} \times \left[\frac{P_W - P_{gi}}{T_W - T_{gi}}\right] \tag{8.9}$$

The total internal heat transfer coefficient is

$$h_{\text{total},w-g} = h_{c,w-g} + h_{r,w-g} + h_{e,w-g} \tag{8.10}$$

The total heat transfer rate is

$$Q_{\text{total},w-g} = Q_{c,w-g} + Q_{r,w-g} + Q_{e,w-g} \tag{8.11}$$

or

$$Q_{\text{total},w-g} = h_{\text{total},w-g} \times \left(T_w - T_{gi}\right) \tag{8.12}$$

Equation (8.13) allows for the calculation of the rate of heat transmission by conduction through the glass cover:

$$Q_{c,gi-go} = \frac{K_g}{Lg}\left(T_{go} - T_{gi}\right) \tag{8.13}$$

8.4.4 Charging mode

Transfer of heat between copper cylinders and the water mass during the charging mode, and the HT-EBE during the hours of sunlight can be written as follows:

$$h_{Lw}A_1(T_L - T_w) + 2I\tau_g\alpha_w A_1 = h_{\text{total}}A_1\left(T_w - T_{gi}\right) + h_{wc}A_2(T_w - T_c) + m_w C_w \frac{dT_w}{dt} \tag{8.14}$$

Equation (8.15) can be used to represent the energy balance between the water mass and the liner of basin:

$$2I\tau_g\tau_w\alpha_L = h_{Lw}(T_L - T_w) \tag{8.15}$$

According to eq. (8.16), the copper cylinder's energy balance equation is as follows:

$$2I\tau_g\alpha_c A_3 + h_{wc}A_2(T_w - T_c) = A_4\frac{K_c}{X_c}(T_c - T_{\text{PCM}}) \tag{8.16}$$

Equation (8.17) can be used to represent the energy balance for PCM:

$$A_4\frac{K_c}{X_c}(T_c - T_{\text{PCM}}) = M_E\frac{dT_{\text{PCM}}}{dt} \tag{8.17}$$

8.4.5 Discharging mode

Equation (8.18) can be used to represent the energy balance equations for a water mass in basin at sunset:

$$h_{wc}A_2(T_c - T_w) = h_{\text{total}}A_1\left(T_w - T_{gi}\right) + m_w C_w\frac{dT_w}{dt} \tag{8.18}$$

Equation (8.19) can be used to represent the energy balance equations for copper cylinder at sunset:

$$h_{wc}A_2(T_w - T_c) = A_4\frac{K_c}{X_c}(T_{\text{PCM}} - T_c) \tag{8.19}$$

Equation (8.20) can be used to represent the energy balance equations for PCM at sunset:

$$A_4\frac{K_{\text{PCM}}}{X_{\text{PCM}}}(T_{\text{PCM}} - T_c) = M_E\frac{dT_{\text{PCM}}}{dt} \tag{8.20}$$

where M_E is the equivalent heat capacity.

There are three conditions:

(i) $M_E = m_{\text{PCM}} C_s$ if $T_{\text{PCM}} < T_m$

(ii) $M_E = m_{\text{PCM}} C_L$ if $T_{\text{PCM}} > T_m$

(iii) $M_E = m_{\text{PCM}} L_s$ if $T_{\text{PCM}} = T_m$

8.5 Results and discussion

The entire experimental process was carried out over several days (from April 26 to May 4, 2017) in order to make sure that the solar intensity values were reasonably close to one another. The temperature fluctuations of the glass, bottom, salty water, and vapor in relation to the time for active SS, when six cylinders are filled with Al_2O_3 at water depth 6 cm in the basin, are shown in Fig. 8.6. The experiment was conducted on April 30, 2022. All temperatures followed a rising trend until they peaked about 2:00 pm, after which they began to decrease gradually until sunset. The maximum value of temperatures of the glass, bottom, salty water, and vapor are 49.4 °C, 53.0 °C, 52.2 °C, and 58.6 °C, respectively.

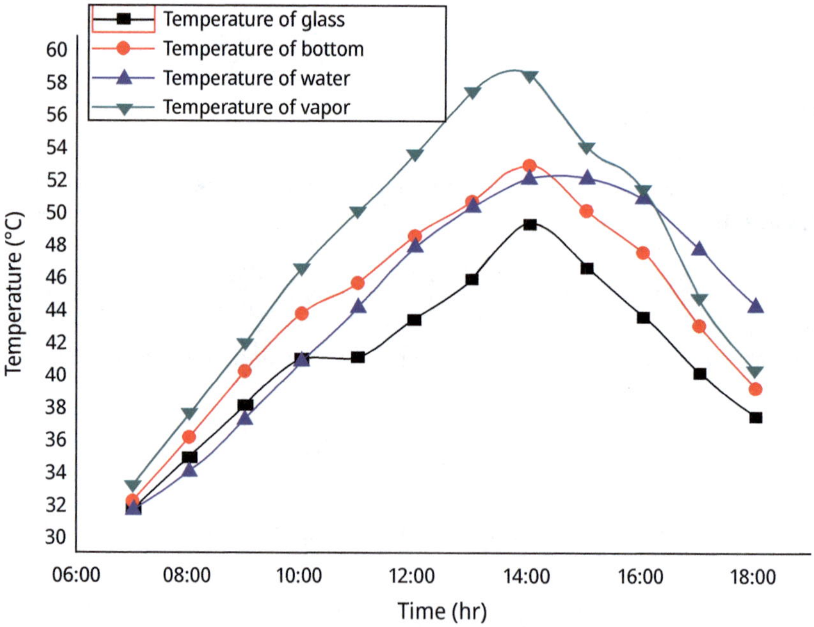

Fig. 8.6: The temperature variations in different segments of the active solar still (all six cylinders are filled with Al_2O_3).

The temperature fluctuations of the glass, bottom, salty water, and vapor in relation to the time for active SS, when six cylinders are filled with paraffin wax at water depth 6 cm in the basin, are shown in Fig. 8.7. The experiment was conducted on April 27, 2022. All temperatures followed a rising trend until they peaked about 2:00 pm, after which they began to decrease gradually until sunset. The maximum value of temperatures of the glass, bottom, salty water, and vapor are 50.1 °C, 55.1 °C, 53.8 °C, and 57.6 °C, respectively.

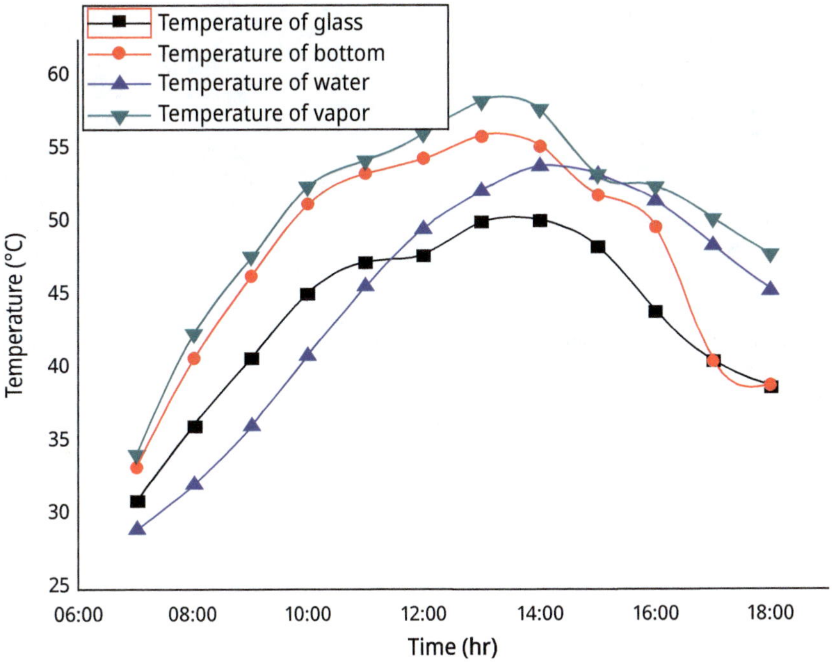

Fig. 8.7: The temperature variations in different segments of the active solar still (all six cylinders are filled with paraffin wax).

The temperature fluctuations of the glass, bottom, salty water, and vapor in relation to the time for active SS, when three cylinders are filled with paraffin wax and three cylinders are filled with Al_2O_3, at water depth 6 cm in the basin, are shown in Fig. 8.8. The experiment was conducted on May 5, 2022. All temperatures followed a rising trend until they peaked about 2:00 pm, after which they began to decrease gradually until sunset. The maximum value of temperatures of the glass, bottom, salty water, and vapor are 45.1 °C, 50.6 °C, 51.6 °C, and 52.2 °C, respectively.

The temperature fluctuations of the glass, bottom, salty water, and vapor in relation to the time for active SS without using PCM at water depth 6 cm in the basin are shown in Fig. 8.9. The experiment was conducted on April 26, 2022. All temperatures followed a rising trend until they peaked about 1:00 pm, after which they began to

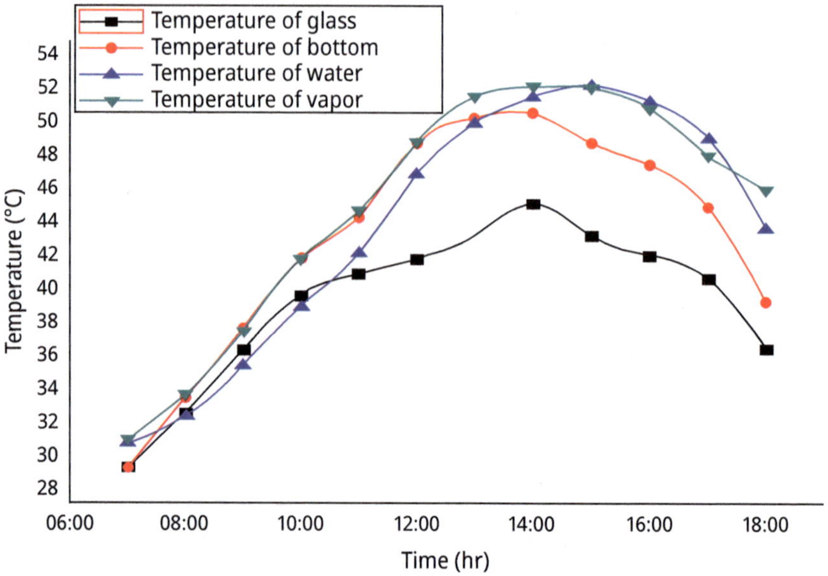

Fig. 8.8: The temperature variations in different segments of the active solar still (three cylinders are filled with paraffin wax and three cylinders are filled with Al_2O_3).

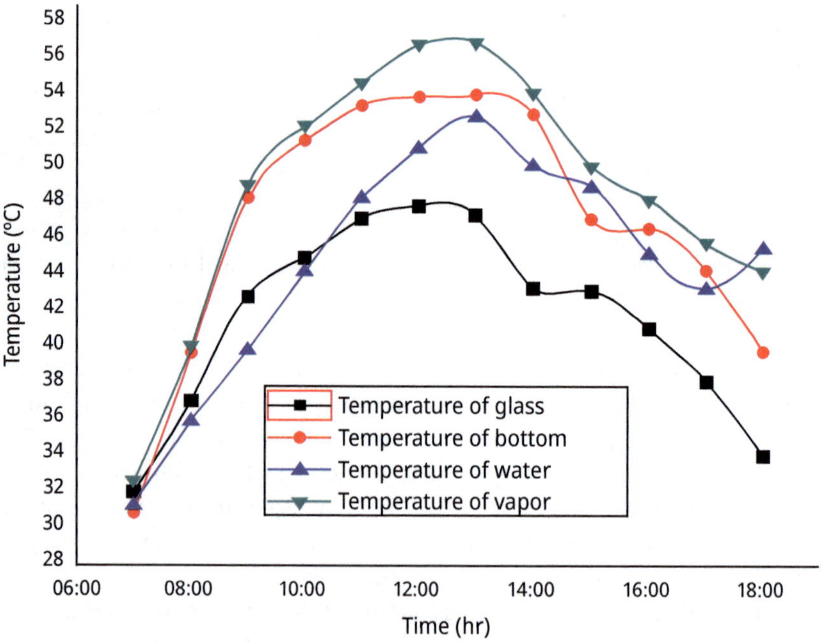

Fig. 8.9: The temperature variations in different segments of the active solar still without using PCM.

decrease gradually until sunset. The maximum value of temperatures of the glass, bottom, salty water, and vapor are 47.4 °C, 54.0 °C, 52.8 °C, and 56.9 °C, respectively.

The temperature fluctuations of the glass, bottom, salty water, and vapor in relation to the time for passive SS, when six cylinders are filled with Al_2O_3 at water depth 6 cm in the basin, are shown in Fig. 8.10. The experiment was conducted on May 1, 2022. All temperatures followed a rising trend until they peaked about 2:00 pm, after which they began to decrease gradually until sunset. The maximum value of temperatures of the glass, bottom, salty water, and vapor are 41.7 °C, 45.9 °C, 48.3 °C, and 49.4 °C, respectively.

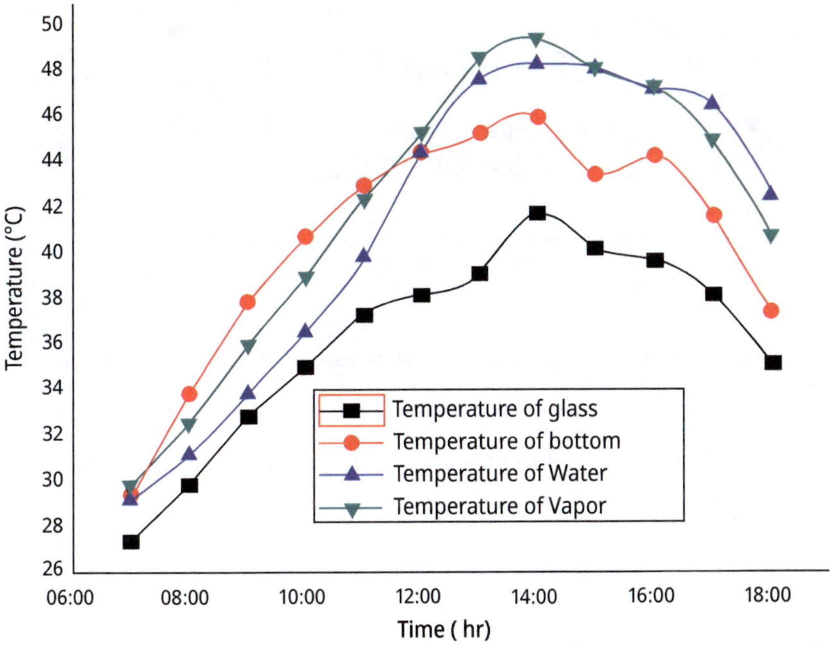

Fig. 8.10: The temperature variations in different segments of the passive solar still (all six cylinders are filled with Al_2O_3).

The temperature fluctuations of the glass, bottom, salty water, and vapor in relation to the time for passive SS, when six cylinders are filled with paraffin wax at water depth 6 cm in the basin, are shown in Fig. 8.11. The experiment was conducted on April 29, 2022. All temperatures followed a rising trend until they peaked about 2:00 pm, after which they began to decrease gradually until sunset. The maximum value of temperatures of the glass, bottom, salty water, and vapor are 52.5 °C, 56.7 °C, 54.9 °C, and 58.5 °C, respectively.

The daily distillates for paraffin wax and Al_2O_3, respectively, were 540 and 510 mL/m^2-day at active modes. The daily distillates for paraffin wax and Al_2O_3, respectively, were 390 and 350 mL/m^2-day at passive modes. Conclusively, it can be seen

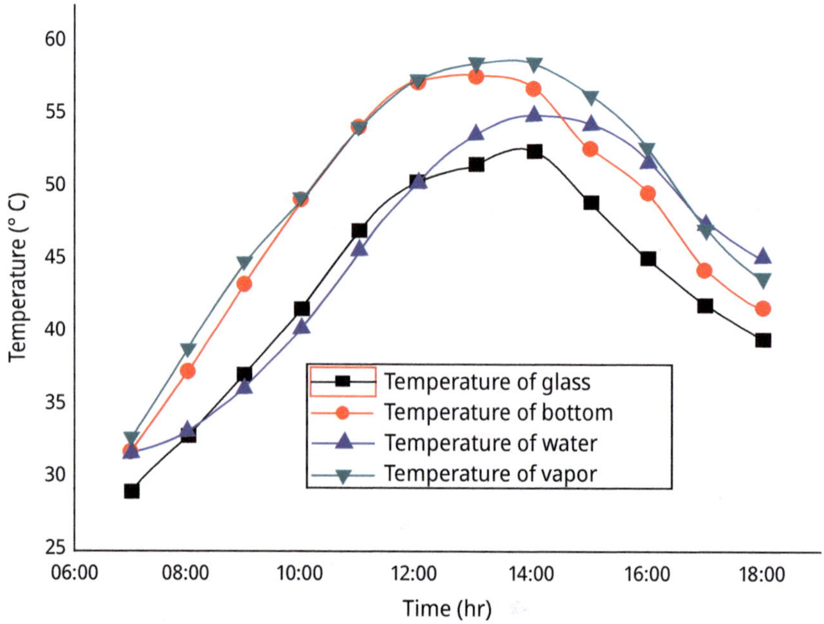

Fig. 8.11: The temperature variations in different segments of the passive solar still (all six cylinders are filled with paraffin wax).

that for all two PCMs, the highest output was 540 mL/m^2-day attained at 6-cm water depth as shown in Fig. 8.12 and Tab. 8.3.

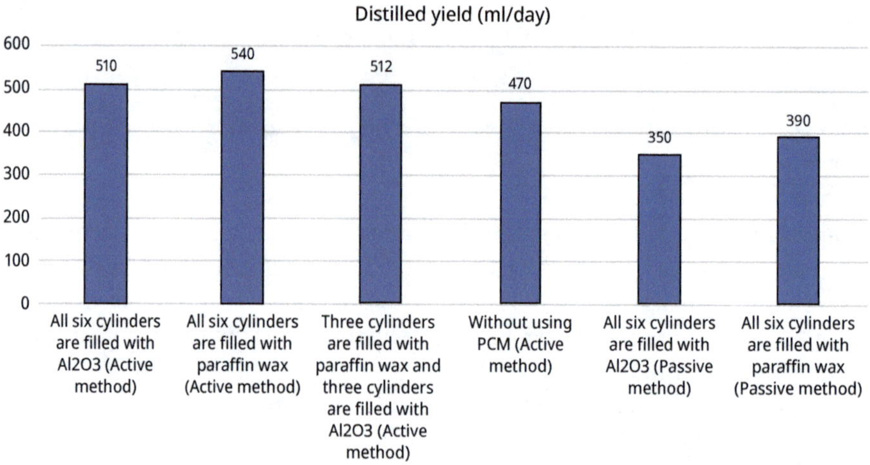

Fig. 8.12: Distilled yield.

8.5.1 Examinations of the distillate water's physical and chemical properties

Authenticity of water generated by SSs is dependent on physical and chemical characteristics. When evaluating whether water is fit for human consumption, these characteristics are particularly crucial. The following comparisons were made between each variable and the authorized parameters mentioned in Tab. 8.4:

- TDS content readings were observed to range at 500 ppm in all water samples that were analyzed.
- Among all the studied samples, the distillate water made using SS produces the best outcomes (31.61 ppm).
- The pH levels are below all of the required ranges for water properties. The finest outcomes were produced by distillate water.
- The quantity and quality of ions in the water determine electrical conductivity, which is expressed digitally as the potential for a water solution to transfer electricity. Water from the SSs has a conductivity of 39.8 μS/cm, which is a significant number. The enhanced water purification system is to blame for this.

Tab. 8.3: Methods of experiments with distillate yield.

S. no.	Conditions	Distilled yield (mL/day)
1	All six cylinders are filled with Al_2O_3 (active method)	510
2	All six cylinders are filled with paraffin wax (active method)	540
3	Three cylinders are filled with paraffin wax and three cylinders are filled with Al_2O_3 (active method)	512
4	Without using PCM (active method)	470
5	All six cylinders are filled with Al_2O_3 (passive method)	350
6	All six cylinders are filled with paraffin wax (passive method)	390

Tab. 8.4: Distilled water examination.

S. no.	Test	Input water	Distilled water	Acceptable limit
1	pH	7.28	7.07	6.5–8.5
2	Total dissolved solids (mg/L)	241.1	31.61	500
3	Turbidity (NTU)	4.91	1.12	1
4	Electrical conductivity (μS/cm)	128.5	39.8	1,000
5	Total hardness	72.3	22.3	200

8.6 Conclusions

One of the best ways to offer freshwater in rural locations with little access to technology and during emergencies would be to use solar water generation. The efficiency of the SS was increased by the integration of copper cylinders containing PCMs above the basin liner. Paraffin wax has produced the highest distillate when compared to Al_2O_3. This is because paraffin wax has a higher latent heat capacity than the other PCMs, according to research. Due to its large surface area, cylindrical storage for PCMs has produced superior results over spherical storage.

References

[1] Pugsley A, Zacharopoulos A, Mondol JD, Smyth M. Global applicability of solar desalination. Renewable Energy. 2016; 88:200–19.

[2] Stuber MD. Optimal design of fossil-solar hybrid thermal desalination for saline agricultural drainage water reuse. Renewable Energy. 2016; 89:552–63.

[3] Qadir M, Sharma BR, Bruggeman A, Choukr-Allah R, Karajeh F. Non-conventional water resources and opportunities for water augmentation to achieve food security in water scarce countries. Agricultural Water Management. 2007; 87(1):2–22.

[4] WHO. World Health Organization, 2004.

[5] Essa FA, Abd Elaziz M, Elsheikh AH. Prediction of power consumption and water productivity of seawater greenhouse system using random vector functional link network integrated with artificial ecosystem-based optimization. Process Safety and Environmental Protection. 2020; 144:322–29.

[6] Younes M, Abdullah A, Essa F, Omara Z, Amro M. Enhancing the wick solar still performance using half barrel and corrugated absorbers. Process Safety and Environmental Protection. 2021; 150:440–52.

[7] Abd Elaziz M, Essa F, Elsheikh AH. Utilization of ensemble random vector functional link network for freshwater prediction of active solar stills with nanoparticles. Sustainable Energy Technologies and Assessments. 2021; 47:101405.

[8] Essa FA, Abd Elaziz M, Elsheikh AH. An enhanced productivity prediction model of active solar still using artificial neural network and Harris Hawks optimizer. Applied Thermal Engineering. 2020; 170:115020.

[9] Essa F, Omara Z, Abdullah A, Shanmugan S, Panchal H, Kabeel A, Sathyamurthy R, Alawee WH, Manokar AM, Elsheikh AH. Wall-suspended trays inside stepped distiller with Al2O3/paraffin wax mixture and vapor suction: Experimental im-plementation. Journal of Energy Storage. 2020; 32:102008.

[10] Essa FA, Omara Z, Abdullah A, Shanmugan S, Panchal H, Kabeel AE, Sathyamurthy R, Athikesavan MM, Elsheikh A, Ab-delgaied M, Saleh B. Augmenting the productivity of stepped distiller by corrugated and curved liners, CuO/paraffin wax, wick, and vapor suctioning. Environmental Science and Pollution Research. 2021; 28(40):955–65.

[11] Essa FA, Abou-Taleb FS, Diab MR. Experimental investigation of vertical solar still with rotating discs. Energy Sources, Part A: Recovery, Utilization, and Environmental Effects. 2021;1–21.

[12] Diab MR, Essa FA, Abou-Taleb FS, Omara ZM. Solar still with rotating parts: A review. Environmental Science and Pollution Research. 2021; 28(39):260–81.

[13] Abdelgaied M, Zakaria Y, Kabeel AE, Essa FA. Improving the tubular solar still performance using square and circular hollow fins with phase change materials. Journal of Energy Storage. 2021; 38:102564.

[14] Arani RP, Sathyamurthy R, Chamkha A, Kabeel AE, Deverajan M, Kamalakannan K, Balasubramanian M, Manokar AM, Essa F, Saravanan A. Effect of fins and silicon dioxide nanoparticle black paint on the absorber plate for augmenting yield from tubular solar still. Environmental Science and Pollution Re-search. 2021; 28(26):102–12.

[15] Kabeel A, Sathyamurthy R, Manokar AM, Sharshir SW, Essa F, Elshiekh AH. Experimental study on tubular solar still using Graphene Oxide Nano particles in Phase Change Material (NPCM's) for fresh water production. Journal of Energy Storage. 2020; 28:101204.

[16] Essa F, Omara Z, Abdullah A, Kabeel A, Abdelaziz G. Enhancing the solar still performance via rotating wick belt and quantum dots nanofluid. Case Studies in Thermal Engineering. 2021; 27:101222.

[17] Kabeel A, Omara ZM, Essa F, Abdullah A, Arunkumar T, Sathyamurthy R. Augmentation of a solar still distillate yield via absorber plate coated with black nanoparticles. Alexandria Engineering Journal. 2017; 56:433–38.

[18] Kabeel AE, Omara Z, Essa F. Numerical investigation of modified solar still using nanofluids and external condenser. Journal of the Taiwan Institute of Chemical Engineers. 2017; 75:77–86.

[19] Panchal H, Nurdiyanto H, Sadasivuni KK, Hishan SS, Essa FA, Khalid M, Dharaskar S, Shanmugan S. Experimental investigation on the yield of solar still using manganese oxide nanoparticles coated absorber. Case Studies in Thermal Engineering. 2021; 25:100905.

[20] Shanmugan S, Essa FA, Gorjian S, Kabeel AE, Sathyamurthy R, Muthu Manokar A. Experimental study on single slope single basin solar still using TiO2 nano layer for natural clean water invention. Journal of Energy Storage. 2020; 30:101522.

[21] Pounraj P, Winston DP, Kabeel A, Kumar BP, Manokar AM, Sathyamurthy R, Christabel SC. Experimental investigation on Peltier based hybrid PV/T active solar still for enhancing the over-all performance. Energy Conversion and Management. 2018; 168:371–81.

[22] Hedayati-Mehdiabadi E, Sarhaddi F, Sobhnamayan F. Exergy performance evaluation of a basin-type double-slope solar still equipped with phase-change material and PV/T collector. Renewable Energy. 2020; 145:2409–25.

[23] Attia MEH, Kabeel AE, Abdelgaied M, Essa FA, Oma-ra ZM. Enhancement of hemispherical solar still productivity using iron, zinc and copper trays. Solar Energy. 2021;216:295–302.

[24] Alaian WM, Elnegiry EA, Hamed AM. Experimental investigation on the performance of solar still augmented with pin-finned wick. Desalination. 2016;379:10–15.

[25] Manokar AM, Winston DP. Experimental analysis of single basin single slope finned acrylic solar still. Materials Today: Proceedings. 2017;4:7234–39.

[26] Omara Z, Hamed MH, Kabeel A. Performance of finned and corrugated absorbers solar stills under Egyptian conditions. Desalination. 2011;277:281–87.

[27] El-Sebaii A. Thermal performance of a triple-basin solar still. Desalination. 2005;174:23–37.

[28] Modi KV, Nayi KH, Sharma SS. Influence of water mass on the performance of spherical basin solar still integrated with parabolic reflector. Groundwater for Sustainable Development. 2020;10:100299.

[29] Nayi KH, Modi KV. Pyramid solar still: A comprehensive review. Renewable and Sustainable Energy Reviews. 2018;81:136–48.

[30] Abdullah AS, Essa FA, Bacha HB, Omara ZM. Improving the trays solar still performance using reflectors and phase change material with nanoparticles. Journal of Energy Storage. 2020;31:101744.

[31] Sharshir SW, Peng G, Wu L, Essa FA, Kabeel AE, Yang N. The effects of flake graphite nanoparticles, phase change material, and film cooling on the solar still performance. Applied Energy. 2017;191:358–66.

[32] El-Sebaii AA, Al-Ghamdi AA, Al-Hazmi FS, Faidah AS. Thermal performance of a single basin solar still with PCM as a storage medium. Applied Energy. 2009;86(7):1187–95.

[33] Singh AK, Tiwari GN. Thermal evaluation of regenerative active solar distillation under thermosyphon mode. Energy Conversion and Management. 1993;34(8):697–706.
[34] El-Samadony YAF, El-Maghlany WM, Kabeel AE. Influence of glass cover inclination angle on radiation heat transfer rate within stepped solar still. Desalination. 2013;384:68–77.
[35] Sonker VK, Chakraborty JP, Sarkar A, Singh RK. Solar distillation using three different phase change materials stored in a copper cylinder. Energy Reports. 2019;5:1532–42.
[36] Singh D, Sharma AK. Experimental performance of modified multi-wick basin-type inverted absorber solar still. Arabian Journal for Science and Engineering. 2021;46(12).
[37] Singh D, Sharma AK. Energy, exergy and enviroeconomic analysis of modified multi-wick basin type inverted absorber solar still. Journal of Mechanical Science and Technology. 2022;36(2):1003–13.
[38] Singh D, Sharma AK. Thermal analysis of modified multi- wick basin type inverted absorber solar still. Journal of Mechanical Science and Technology. 2021;35(5):2271–82.
[39] Dunkle RV. International development in heat transfer. In: ASME Proceedings. Inter. Heat Transfer. University of Colorado, 1961, 895. Part V
[40] Chen Z, Yao Y, Zheng Z, Zheng H, Yi Yang LH, Chena G. Analysis of the characteristics of heat and mass transfer of a three-effect tubular solar still and experimental research. Desalination. 2013;330:42–48.

Rakesh Pant, Khushi Sah, Nirmal Patrick, Amit Gupta

9 Role of biomass, agriculture waste, and municipal waste in production of bio-based lubricants

9.1 Introduction

The world population growth, combined with industry and technology, has resulted in increased energy demand. Several renewable energy sources have been presented as possible energy sources, including hydropower, geothermal, wind, sun, and biomass energy. Among them, biomass is the sole renewable source of energy and chemical compounds, making it the only contemporary option capable of replacing petrol in the fusion of a wide variety of value organic goods. Lubricants are said to be produced in amounts ranging from 30 to 40 million tons per year; these compounds are frequently employed in a variety of industrial applications to reduce friction and heat, protect from corrosion and wear, convey energy, remove impurities or sealing methods, amid other uses [1]. Biolubricants are any lubricants that easily decompose and pose no threat to marine life or people. They are also referred to as viable, recyclable, or ecologically sound lubricants. It is believed that around half of all oil consumed ends up in the ecosystem. This creates a significant encouragement to employ recyclable lubricants.

Based on these challenges, worries about the environment, and viability, the biolubricant sector is expanding. These are products that are made from biodegradable and renewable raw resources. This definition, however, is not commonly recognized. In other cases, only biodegradability is taken into account in the definition. Biolubricants will be described as decomposable, renewable materials for now. Biolubricants do not have to be totally made from vegetable oil base stocks [2]. Huge volumes of rubbish are piling up all around the world at an escalating rate, yearly. In 2016, 2.01 billion tons of solid garbage was produced, resulting in a daily footprint of 0.79 kg per person. In India alone, the local, farming, manufacturing, mineral extraction, and additional fields create over 960 million tons of solid garbage every year. Out of 960 million tons of solid waste, 360 million are organic leftover from agronomy, 290 million tons are inorganic leftover from the industrial and mining industries, 4.5 million tons are. Agricultural waste is biodegradable and environmentally beneficial, although the breakdown process can be slow and time-consuming in some circumstances. Farmers aim to clear their fields of preceding harvest waste as quickly as possible in order to prepare them for the next crop season. As other waste management options are expensive, many farmers choose to burn crop wastes in open areas, which pollute the ecosystem. It is quickly becoming critical to create more sustainable and environmentally friendly waste management options [3].

https://doi.org/10.1515/9783111145136-009

Plants have long been used to create biomaterials and biofuels. During the twentieth century, the creation of oil from fossil fuels superseded many of these traditional applications. The possibility of depleting oil supplies and concern about how including carbon in fossil fuels may affect the environment has renewed interest in using flora as a direct source of bioenergy and biomaterials [4]. Currently, many materials have been employed as raw materials in scientific research, like wood, leaf wood, industry leftovers, rice husks, sawdust, and maize stalks. Bio-oil's heat value decreases and its preservation durability is affected by high water content [5].

Renewable fuels, on the other hand, are gaining popularity as a means of replacing traditional fuel and filling the energy deficit. Since it may create fuel with properties similar to those of crude oil, biofuel is one of the possible sources of green energy that has gained appeal across the globe. Biofuels are manufactured from biomass using thermochemical or biochemical transformation methods [6]. Creating bioproducts from farming leftovers has the potential to be very profitable, but technologically speaking, methods for converting biomass require improvements and a better understanding of their capabilities, along with economic, ecological, and societal sustainability. Despite mounting environmental concerns (e.g., global warming, the seventh plastic continent), the most extensively used chemicals, adaptable plastics, and energy are still provided by the petrochemical industry. Nearly every chemical and component for plastics may be produced using raw materials that are recyclable, according to chemical and biotechnological experts [7].

From the current limit of 3%, worldwide biofuel supply is expected to satisfy 8% of total road transport fuel requirements in 2035. Fuel oxygenates, fuel ethers, and other bio-ethers are used as octane boosters. They reduce pollution emissions; this raises the quality of gasoline. Biogas is the gas created by the breakdown of CO, H, and trace amounts of CO_2 [8].

The lubricant works as an antifriction medium, allowing for smooth operation, ensuring dependable machine operations, and lowering the danger of frequent failure. Lubricants can be liquids, semisolids, or solids. The primary goals of lubrication include:
- Decreasing wear and avoiding heat loss owing to the contact of moving surfaces
- Decreasing oxidation to protect the surface from corrosion
- In transformer applications, it acts as an insulator
- Functioning as a dust, grime, and water repellent

While lubricants cannot completely prevent wear and heat, they can be minimized and managed to acceptable levels. The following are the three primary roles of lubricants:
- Managing friction
- Contact with a cleaning agent
- Cooling the contact

The industrial and transportation industries rely heavily on fossil fuel-based goods like petroleum products and petrochemicals. These goods can pose major environmental risks and pollute the environment. Biomass and solid wastes have the ability to provide fuels for power generation, including platform chemicals, methanol bio-oil, and liquid fuels [9].

9.1.1 Overview of biomass as a rentable and lubricants resource

The material produced by plants known as biomass is created when sunlight, water, and atmospheric CO_2 combine to make the components of biomass, called "carbohydrates." A fraction of 1% of the solar energy that is accessible is normally converted into chemical energy during photosynthesis. The chemical bonds that make up the structural constituents of biomass store the sun's energy that powers photosynthesis. When biomass is effectively broken down chemically or biologically, the energy contained in chemical bonds is extracted, and the resulting "energy" product combined with oxygen causes the carbon to be oxidized to generate carbon dioxide and water. As CO_2 is once again available for the creation of new biomass, the process is cyclical [10]. A nonfossilized and recyclable organic material called "biomass" is made via flora, fauna, and bacteria. Additionally, biomass consists of the nonfossilized and decomposing organic components of industrial use and waste from municipalities, plus products, by-products, remains, and rubbish from related industries including forestry and agriculture. Gases and liquids that are produced when nonfossilized and compostable organic materials break down are referred to as "biomass" [11].

The main source of feed in another facility was biological trash from the first. Biorenewable resources include agricultural waste, yard and garden debris, municipality solid trash, culinary junk, animal manure, and other waste materials. The capacity of existing landfills has been reached.

Current energy supplies have finite reserves and are depleting as the world populace grows and technology advances. Biomass resources are plentiful all around the planet. Locally, there is a poor level of utilization of biomass, especially leftover biomass from agriculture, and the unneeded residue has been disposed of in ecologically unfavorable ways, such as by burning it in fields and creating carbon dioxide-filled black clouds. The characteristics of biomass in terms of rejuvenation and abundance make it an appealing source of renewable energy.

Biomass power "bioenergy" is a viable approach for manufacturing biofuel and energy from plants since it has a recycling array of corn, which is regarded as a vital food in numerous regions of the globe; the yearly worldwide output is 864.96 million metric tons. These agronomies generate around 1,730 million metric tons of residues (leaves, stalks, and straw) every year. The entire number of date palm trees in the globe is estimated to be over 100 million. Egypt's entire amount of agricultural waste from maize is roughly 7.3 million tons of dry matter yearly; it has approximately

16 million date palm trees. Under typical development circumstances, each date palm produces an average of 12–15 new leaves [12].

9.2 Resource of biomass

Researchers characterize biomass in a variety of ways, but one basic strategy is to divide it into four categories: woody floras, herbaceous floras/grasses, marine floras, and composts. Paper, textiles, and wool products, as well as food, yard, and wood waste, are examples of biogenic materials in municipal solid waste [13]. Biogas is produced from animal waste and human sewage. Multiple techniques are used to convert biomass to energy, such as:

- Direct combustion (burning) to create heat
- Thermochemical synthesis of solid, gaseous, and liquid fuels
- Liquid fuels are created by chemical conversion
- The use of biological processes to create liquid and gaseous fuels

9.3 Conversion technologies for biomass lubricants

9.3.1 Thermochemical conversion

Combustion, gasification, pyrolysis, and liquefaction are the main thermochemical methods of transformation of biomass. The thermochemical transformation of biomass that is most frequently used in industry is combustion, which is used to generate heat and electricity [13]. Thermochemical transformation requires bond-breaking and then restructuring of organic materials into charcoal, synthesis gas, and highly oxygenated bio-oil at high temperatures. The choice of the transformation type can be influenced by the kind and quantity of biomass feedstock and the required type of energy – for instance, end-use situations, ecological considerations, financial circumstances, and project-specific variables. Therefore, in order to evaluate the potential of the numerous thermochemical procedure options, comprehending them is essential (Figs. 9.1 and 9.2) [14].

9.3.2 Gasification

Flames are produced as a result of this process's thermal breakdown of the solid phase's outer surface, which releases volatile and combustible materials that join thermal reactions in the gas phase [15]. The creation of a combination of flammable gases that might involve H_2, CH_4, CO, and CO_2, requires the employment of gasifying

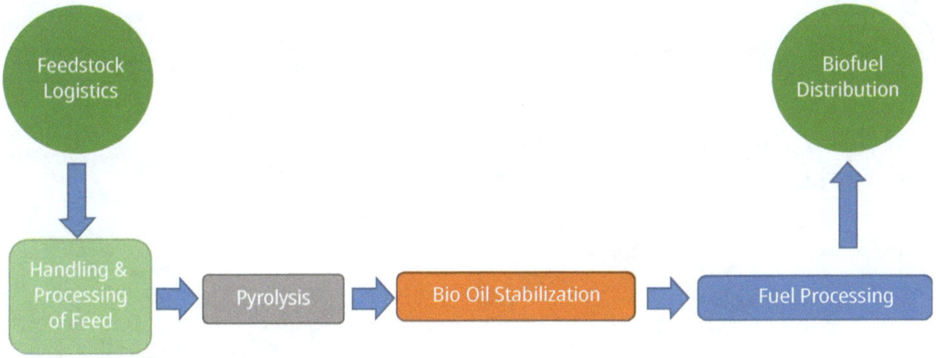

Fig. 9.1: Thermal platform integration pyrolysis.

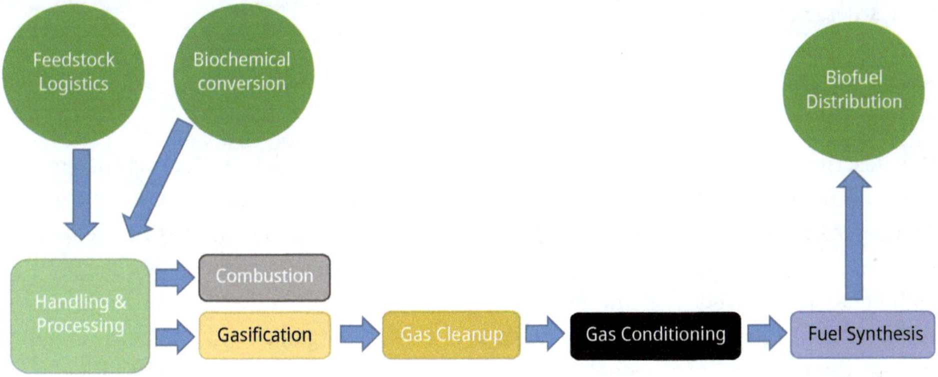

Fig. 9.2: Thermal platform integration gasification.

agents like air, steam, or CO_2. Temperatures between 800 and 1,300 °C are used for the operation. Currently, biomass gasification might be cohesive with a variety of industrial operations and power generation systems due to the elasticity of the gasification technology and the numerous uses of the generated syngas [16]. The function of the gasifier is based on the moisture content of the biomass feed used. This is due to the fact that the method uses more energy or heat to vaporize the additional moisture to syngas temperature (Fig. 9.3).

Gasification produces condensable liquids rich in water and biochemical in addition to the major gaseous components. Due to the greater temperature, tar development is limited, making cleaning and recovering the gas relatively straightforward. In addition, high temperatures in gasification speed up the reaction rate, which is favored by endothermic reactions including methanation, water-gas shift, and steam reforming. This causes the breakdown of biomass to gases to be almost entirely complete. Syngas may be converted through Fischer-Tropsch catalysis into clean fuels and value-added products [17].

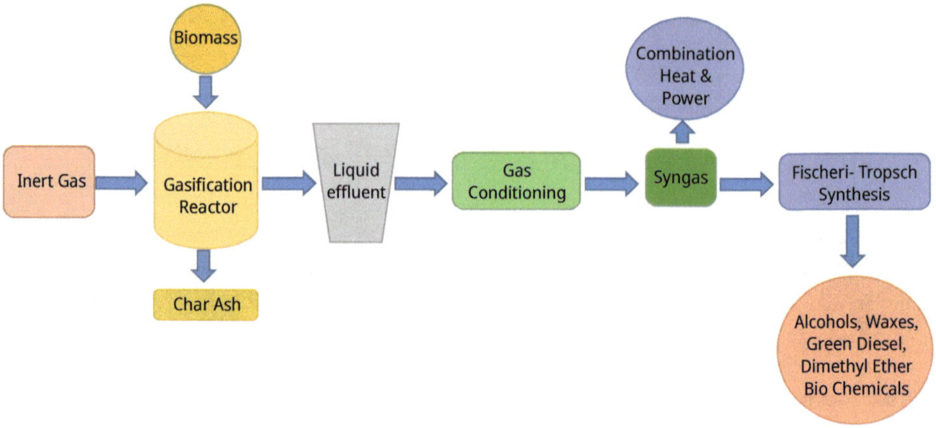

Fig. 9.3: Process of gasification of biomass.

9.3.3 Pyrolysis

It is described as the heat depolymerization of organic materials in the presence or absence of nitrogen. Pyrolysis is an exothermic process that produces vapors and carbon-rich remains with heat needs ranging from 207 to 434 kJ/kg. Bio-oil may be created by condensing these vapors ($C_6H_{10}O_5$, $C_5H_{10}O_5$, $C_{81}H_{92}O_{28}$). But the carbon residues are left as biochar [18]. Pyrolysis is the practice of heating organic material at high temperatures without O_2. The absence of oxygen prevents the organic material from burning. Rather, the chemical components of the substance break down into combustible gases and charcoal.

Char's properties are determined by the pyrolysis conditions. Char's hydrogen and oxygen levels both drop as the temperature rises. The H/C ratio drops from 0.49 at 400 °C to 0.15 at 700 °C, showing that the aromaticity and carbonaceous character of char increases [19]. Pyrolysis systems are classified into two categories. Fast pyrolysis increases the rate of pyrolysis temperature to 1,000 °C/s, which optimizes bio-oil production. This method produces 60–70% bio-oil, 15–25% biochar, and 10–15% syngas. Slow pyrolysis, on the other hand, employs slower heating rates and the primary end product is biochar. The system is self-sustaining in both processes since the syngas is used to fuel the reaction [20]. Conventional fast pyrolysis involves rapidly heating biomass feedstock to roughly 932 °F (500 °C) at atmospheric pressure and without the use of a catalyst in less than two seconds. Pyrolysis vapors are quickly extinguished and caught [21]. The main components of cooled pyrolysis products include liquid (water and organic chemicals), char mixed with biomass ash, and non-condensable gases. Traditional quick pyrolysis methods are already in use for the manufacture of food flavorings and heat/power applications (Fig. 9.4).

Ablative pyrolysis, in contrast to conventional fast pyrolysis of biomass, may employ larger pieces of biomass as feedstock rather than only tiny particles, reducing the need for grinding. Typically, ablative pyrolysis involves big chunks of biomass that go through a sequence of melting and/or sublimation processes [22].

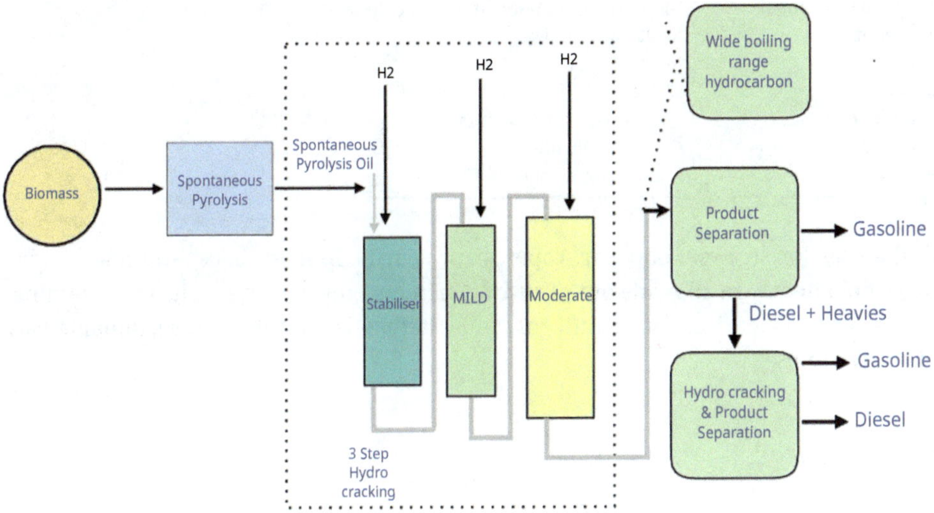

Fig. 9.4: Flow diagram of pyrolysis.

9.4 Utilizing agricultural waste for bio-based lubricants

9.4.1 Straw

The stalks of cereal plants are used to make straw once there has been a separation of grain from chaff. The yield of this crop is usually around half. In agricultural areas and farms, straw is gathered, stacked in bales or bundles, and then kept for later use after being expertly tied with a sturdy rope. Rice, corn, wheat, oats, barley, soybeans, millet, and a variety of grains are used to make straws. Approximately 800–1,000 million tons (MT) of rice straw are produced yearly. This is apart from its transformation into biogas, with around 600–800 MT coming from Asia. The percentages of cellulose, hemicellulose, and lignin in rice straw are 32–47%, 19–27%, and 5–24%, respectively [15, 16]. The waste obtained from different crops is shown in Tab. 9.1.

Additionally, it is predicted that 354 MT and 203 MT of wheat straw and maize straw, respectively, are produced yearly on a worldwide scale. Based on recent data, oat, barley and millet are respectively produced in 29.5, 23, and 142 MT per year in the world. The building industry, horticulture, packaging, and animal feed are just a few

Tab. 9.1: Waste obtained from different crops, by products.

# Crop	Waste
1 Rice	Bran, straw, stalks, hull, husk
2 Sugarcane	Bagasse
3 Agriculture crops	Mixed crops and not limited to crop waste
4 Peanuts	Shells
5 Nuts	hulls
6 Cotton	stalks
7 Corn	Cob, stover, stalks, leaves
8 Coffee	Ground, hull, husk
9 Coconut	Fronds, husk, shell

of the uses for straws. Footwear, ropes, papers, hats, baskets, beds, and many more household items are all made out of straw. Open burning, which results in contamination, a high risk of fire, and significant deforestation, is a farmers' unsustainable way of straw disposal.

9.4.2 Husk

The fibrous outer layer of a seed or grain is called husk, commonly referred to as the hull. In order to gain access to the grain/seed, the husks of grains/seeds are typically peeled while processing. Among various things, husks are present in corn, hazelnut, coffee, peanut, chocolate, barley, rice, cowpea, soybeans, and peanut. Population expansion has boosted food production, which means that more husks from rice, corn, cowpeas, barley, soybeans, and other food grains are produced. Husks can be used to make fuel, plant-based food, insulating materials like particle boards, and cardboard.

9.4.3 Bagasse

It is a diverse, multicellular, lignocellulosic residue fiber that is left over after the juice or sugar has been taken from a fruit/ seed/ vegetable stalk. Bagasse is the by-product produced when juice is extracted from sugarcane, sorghum, cashew, and pomegranate stalks. Approximately 279 MMT of sugarcane bagasse was produced in 2021 from 1.6 billion tons of sugarcane that was cultivated. Bricks, ceramics, concrete, cement additives, adsorbents, heterogeneous catalysts, and biofuels are all made from sugarcane bagasse. A cheap feedstock for the production of biofuel, chemicals, bio-products, and animal feed supplements is cashew apple bagasse.

9.4.4 Peels

The thick skin, covering, or rind of a fruit and vegetable is referred to as a "peel." To get to the fruit or vegetable, the covering or skin must be peeled off. Though peels are inedible and occasionally harmful, they are utilized for a variety of purposes. Some peels, such as orange, banana, plantain, pomegranate, and others, such as cucumber, yam, cassava, potato, ginger, and others, are difficult to separate or detach. Banana production increased to about 119.8 MT in 2020 from 110 MT in 2016. Considering that the peel makes up at least 35% of the weight of the entire banana fruit, the banana peel production for 2020 was 42 MT. With a 359 MMT global potato production in 2020, potato peel waste amounted to 54–143 MMT because 15–40% of the potato tuber was removed as peel [2–4]. They have several uses in the home, in body care, nutraceutical industry, as well as biotechnology. For instance, manufacturing of biofuels, bioplastics, biocomposites, nanocrystals, lactic acid, and extra by-products all use potato peels as feedstock [8].

9.4.5 Shells

It is the solid outer covering that protects and covers an egg, seed, or nut. The shells must be cracked to obtain access to the seed or nut. Groundnut, palm kernel, and egg shells all include shells. A palm kernel shell is a crop remnant produced by shattering the kernel and extracting the seed contained within the shell. The eggshell serves as an exterior protective coating for the egg, protecting it from cracking, injury, and germ assault. Annually, around 38.6 MT of groundnut are produced globally, with the shell accounting for roughly 20% of the total weight. It implies that a massive volume of groundnut shell residue is produced each year. The 25 MT of waste eggshell is created each year, making chicken eggshell as the worlds an environmental and food resource for the fifteenth-century contamination [20]. When eggshells are left in landfills, they produce an awful stench and serve as breeding ground for mosquitos and other insects. The massive shells produced by palm kernel processing have found use as adsorbents for effluent treatment and reinforcing for Portland cement [23].

9.5 Extraction techniques for obtaining valuable components from agricultural waste

In 2009, there were 3.6 million tons of dissolving pulp manufactured worldwide. Between 1998 and 2009, global dissolving pulp production increased by 24%. Manufacturing has risen all through the world, with Asian manufacturing increasing by 27% and African manufacturing by 19%, while manufacturing in the Americas climbed by just over 1%. Annually, cellulose products alone create 4.4 billion tons of solid garbage in Asia, with MSW accounting for 790 million tons, of which around 6% metric tons are produced in India. Each year, agronomic resources produce more than 350 MT of garbage. However, it has been stated that over 600 MT of trash has been created in India alone from farming sources. Bagasse, straw and husk, waste fruits and vegetables, foodstuffs like coffee and tea, oil cakes, nutshell, and cotton stalk are the most important agricultural wastes [24, 25].

9.6 Solvent extraction

It is the predominant method of dissolving oil by contacting oilseeds with a liquid solvent. This is most effective way to extract oil from oilseeds. The setup of the oilseeds prior to extraction, the temperature, the process method, and the equipment designs all influence efficiency. The remaining oil in the meal is expected to be below 1% following commercial solvent extraction. The oilseed processing sector is under pressure to use environmentally friendly solvents such as alcohol or water. The principal drawbacks of utilizing alcohols and water for oil extraction are low oil solubility and increased energy needs for solvent recovery and meal drying. Additionally, alcohols reduce the functioning of soybean proteins [20–25]. The extraction temperature enhances the solubility of oil in a solvent. High temperatures also improve the viscosity and diffusivity of oil (Fig. 9.5).

Extraction of oil using the small batch techniques was widespread the European Union in the 1800s. Both extraction and meal desolventizing were carried out in the same vessel. Batch extractors are now used for the process of obtaining specialty and valuable oils, and the waste bleaching clay can be used to recover oil. Two typical batch-type extractor designs are a revolving with an agitator and a vapor-tight lid, a drum and a vertical cylindrical kettle are used.

Continual processes at large scales were first used a century ago. Flaked oilseeds are totally immersed in a solvent in immersion extractors. To obtain optimum contact between seed and solvent in such a system, particle size should be as tiny as feasible. The primary downside of this design is the difficulty in separating solids from miscella (oil + solvent combination). As a result, a type of extractor that uses immersion is appropriate for small-scale manufacturing operations. In extractors of the percolation type, flakes of seeds are molded onto a fixed bed and sprinkled with solvent. The oil is removed from the bed by the solvent as it drains. A suitable flow rate of solvent across the seedbed requires a large enough seed particle size. The bed's fine particles impede the solvent's flow, causing flooding and decreasing the effectiveness of extraction. Initial percolation-based designs consisted of perforated-bottomed basket-style extractors that were filled with flaking seeds. These gadgets have a bucket-lift feel to them [24].

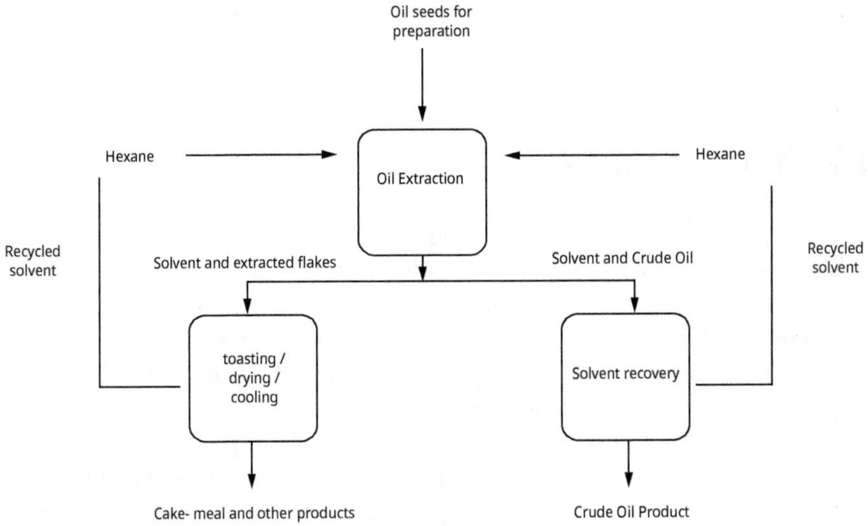

Fig. 9.5: Solvent extraction method of oil.

Flaked seeds are poured into every basket from an automatic feed container at the top. The baskets start to drop, and a solvent is sprayed over them. On the other side of the feed container, the wasted flakes in baskets ascend to the top of the housing; eliminated seeds are poured into a container and then transported to a meal desolventizer via a conveyor belt after being automatically reversed at the top of the baskets. Basket extractors are huge and challenging to upkeep. Rotational and straight extractor designs are the most recent. The baskets in horizontal extractors spin in a single horizontal plane as opposed to vertically as they do in basket-type extractors. On different solvents, extractor designs are discussed in the parts that follow [18].

9.7 Meal and oil desolventizing

The leftover solvent in a meal is eliminated after oil extraction in a desolventizing-toaster (DT) that can have three portions for desolventizing, desolventizing, and toasting. The food is cooked to a high temperature to vaporize the solvent while maintaining nutritional content. DT is made up of trays that are positioned on top of each other along a vertical shaft and have sweep arms connected to move the food around. DT meals may still include moisture, which must be eliminated.

Distillation columns are used to extract solvent from miscella. A second stripping column, eliminates the remaining moisture and solvent from the oil. Condensed solvent vapors from the distillation and stripping columns are returned to the extractor.

After stripping, oil is refined, which eliminates undesired components from the oil [15].

9.8 Mechanical oil extraction

Lever and wedge presses were the earliest pieces of mechanical oil expression machinery used in the processing of oilseeds. The first cottonseed oil mill constructed in the US in the 1920s made use of a hydraulic press. Seeds in filter bags were traditionally inserted into perforated horizontal chambers between the press's head block and ram. The boxes were pressed together by increasing the ram's hydraulic pressure. The filter bag's oil was forced out. Next, the hydraulic presses used a cake filter bag that was extracted by hand. Cages were used in place of filter bags in later types of hydraulic presses. Continuous screw presses and solvent extraction equipment.

The present-day olive oil industry makes use of the hydraulic press. In an automated screw press, pressure is applied to the oilseeds as they go down the press's length by a straight screw with an increasing body diameter and an upright feeder. The barrel that surrounds the screw has grooves across its length that enable the building internal pressure to first release air and then drain the oil through the barrel. The primary advantage of the screw press is that it allows for continuous oil extraction while handling enormous quantities of oilseeds with little labor. Screw presses need challenging maintenance. Wear is a result of the moving parts as well as the elevated pressure and temperature produced by the press. The discovery of construction materials, which increase the life of the parts that wear out the most, has been the main innovation in mechanical screw press design. Previously, parts would need to be changed every three months and now they may last up to 2 years. There are now much larger screw presses available. The daily capacity for complete pressing has risen from 5 tons to over 100 tons, and for pre-pressing applications, to over 800 tons [23, 24].

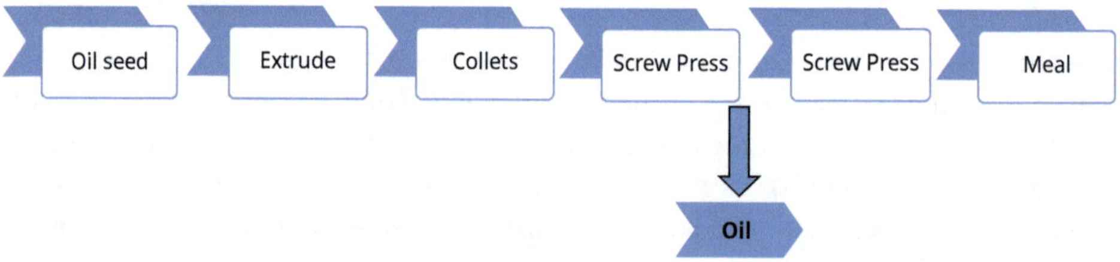

Fig. 9.6: Mechanical oil extraction.

Farmer-owned cooperatives and on-farm businesses have been formed to process locally grown soybeans or other oilseeds using extruding-expelling (E-E) micro-crushing mills (Fig. 9.6). These mills use electric heat instead of steam heating that would require steam-generating equipment. A dry extruder creates heat through friction prior to screw pressing. A partially defatted meal having a remaining oil level of 6–7% is used in cow diets.

Collets have a higher bulk density than oilseeds; hence less space is needed in the extractor for a given weight of oilseeds. The capacity of a plant rises when collets are used for solvent extraction rather than flaking seeds. High oil-content seeds, like rapeseed, sunflower, safflower, peanut, and canola, produce too much oil to manufacture useful flakes and collets.

In order to prepare them for solvent extraction, they are pre-pressed to reduce the meal oil content to 16–18%. Since soybean has a low oil content (18–20%), pre-pressing is not necessary. Pre-pressing and the production of extruded collets are combined into one processing unit with the help of the revolutionary screw press design known as Hivex.

9.9 Aqueous extraction

Expanders can be used to transform oil seeds into collets, porous pellets. The effectiveness of solvent extraction is increased by converting broken seeds into extruded collets. In comparison to flakes, collets are bigger, stronger, and have a higher bulk density. When collections are utilized for solvent extraction, the solvent extractor generates increased solvent flow and micelle (oil + solvent) drainage. Collets fit into the extractor more compactly than an equivalent weight of oilseed because they have a higher bulk density. The manufacturing potential of a plant increases when collets are employed for solvent extraction rather than flaking seeds. Sunflower, safflower, peanut, canola, and rapeseed oily seeds all generate far too much oil to make useful flakes and collets. Due to its low oil content (18–20%), soybeans need not be pre-pressed. Pre-pressing and extruded collet manufactures are combined into one processing unit with Hivex, a groundbreaking screw press design. An expander eliminates the need for a separate extruder and pre-press by incorporating a drainage cage and pressing screw within its barrel.

9.10 Supercritical fluid technology

The search for substitute solvents for the processing of edible oils has been sped up by worries about the risks of organic solvents to the ecosystem and human wellness, as well as government monitoring. Supercritical fluids (SCFs), ethanol, isopropyl alcohol, acetone, and iso-hexane are a few of the substitute solvents that have been studied. Oil and oilseed processing methods have been studied in SCF for more than

20 years as an alternative. Generally, supercritical carbon dioxide (SC-CO_2) is used as a solvent in culinary applications, which is at temperatures and pressures above its critical point.

The capital expense of high-pressure vessels continues to impede the economic viability of SC-CO_2-based technologies. However, cost savings can be gained by simplifying separations and using SCFs with reduced solvent costs. The nontoxicity and drastically decreased waste streams of SC-CO_2 make this technology unique in terms of fulfilling increasingly strict health standards (Fig. 9.7).

The simplicity with which the solvent may be extracted from the extract makes SC-CO_2 stand out from other compounds. When the pressure in the system is released, oil separates from the CO_2-oil combination and carbon dioxide transitions back to the gas phase. Since carbon dioxide is recycled, the system's CO_2 emissions do not affect the ecosystem. The bulk of current applications for SC-CO_2 technology involve high-value products, despite the fact that industrial use has been steadily growing. The economic feasibility of SC-CO_2-based solutions continues to be constrained by the capital cost of high-pressure containers. Simplifying separations and using SCFs with cheaper solvents can, nevertheless, result in cost reductions. In terms of meeting increasingly stringent health requirements, SC-CO_2 is unique due to its nontoxicity and significantly reduced waste streams [25, 26].

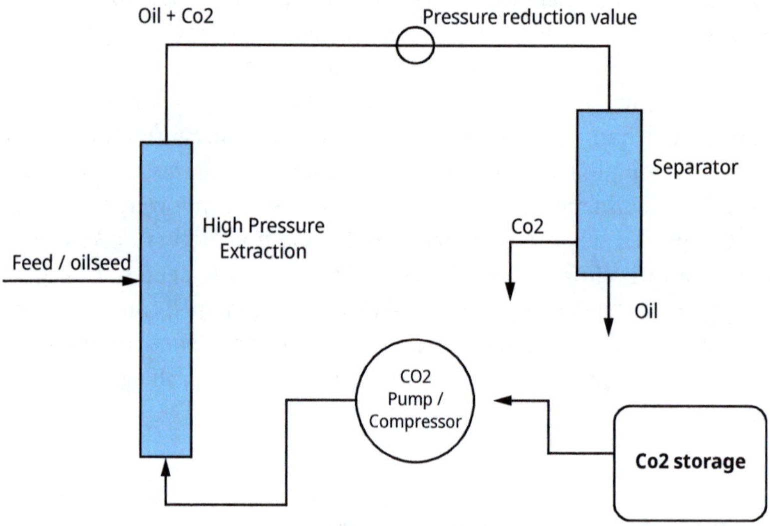

Fig. 9.7: Flow diagram of supercritical carbon dioxide extraction system.

References

[1] Mohamed OA. Production of bio-oil from agriculture waste. Biochemistry Analytical Biochemistry. 2018;7(1):1–9.
[2] Cecilia JA, Plata DB, Saboya RMA, Luna FMTD, Cavalcante Jr CL, Enrique Rodríguez-Castellón ER. An overview of the bio lubricant production process: Challenges and future perspectives. Processes. 2020;8(257):1–24.
[3] Bremmer BJ, Dr.Plonsker L Bio- based lubricants a market opportunity study update. United Soybean Board. 2008;1–31.
[4] Sharma G, Kaur M, Punj S, Singh K. Biomass as a sustainable resource for value-added modern materials: A review. Biofuels, Bioproducts and Biorefining. 2020;1–23.
[5] Henry RJ. Evaluation of plant biomass resources available for replacement of fossil oil. Plant Biotechnology Journal. 2010;8:288–293.
[6] Xu Y, Hu X, Li W, Shi Y. Preparation and characterization of Bio-Oil from biomass. Progress in Biomass and Bioenergy Production. 2011;197–222.
[7] Panwar NL, Paul AS. An overview of recent development in bio-oil upgrading and separation techniques. Environmental Engineering Research. 2021;26(5):1–17.
[8] Gontard N, Sonesson U, Birkved M, Majone M, Bolzonella D, Celli A, Angellier-Coussy H, Jang GW, Verniquet A, Broeze J, Schaer B, Batista AP, Sebok A. A research challenge vision regarding management of agricultural waste in a circular bio-based economy. Critical Reviews in Environmental Science and Technology. 2018;48(6):614–654.
[9] Afolalu SA, Salawu EY, Ogedengbe TS, Joseph OO, Okwilagwe O, Emetere ME, Yusuf OO, Noiki1 AA, Akinlabi SA. Bio- agro waste valorization and its sustainability in the industry: A review. Materials Science and Engineering. 2021;1–12.
[10] Sharma G, Kaur M, Punj S, Singh K. Biomass as a sustainable resource for value-added modern materials: A review. Biofuels, Bioproducts and Biorefining. 2020;1–23.
[11] Ahmad SFK, Ali UFM, Isa KM. Compilation of liquefaction and pyrolysis method used for bio-oil production from various biomass: A review. Environmental Engineering Research. 2020;25(1):18–28.
[12] Sarma RN, Vinu R. Current status and future prospects of bio lubricants: Properties and applications. Lubricants. 2022;10(70):1–21.
[13] Garba A. Biomass conversion technologies for bioenergy generation: An introduction. Biotechnological Applications of Biomass. 2020;1–16.
[14] Lee SY, Sankaran R, Chew KW, Tan CH, Krishnamoorthy R, Chu DT, Show PL. Waste to bioenergy: A review on the recent conversion technologies. BMC Energy. 2019;1(4):1–22.
[15] Mai TP, Nguyen DQ. Gasification of biomass. Gasification of Biomass. 2020;1–13.
[16] Garba A. Biomass conversion technologies for bioenergy generation: An introduction. Biotechnological Applications of Biomass. 2020;1–16.
[17] Jha S, Nanda S, Acharya B, Dalai AK. A review of thermochemical conversion of waste biomass to biofuels. Energies. 2022;15(6352):1–23.
[18] Garba A. Biomass conversion technologies for bioenergy generation: An introduction. Biotechnological Applications of Biomass. 2020;1–16.
[19] Sukumar V, Manieniyan V, Sivaprakasam S. Bio oil production from biomass using pyrolysis and upgrading – a review. International Journal of Chemtech Research. 2015;8(1):196–206.
[20] Warintorn B, Channarong A, Malee S. Conversion of waste cooking oil into bio-fuel via pyrolysis using activated carbon as a catalyst. Molecules. 2023;28(3590):1–18.
[21] Jones SB, Snowden-Swan LJ, Meyer PA, Zacher, Olarte MV, Drennan C. Fast Pyrolysis and Hydrotreating: 2014 State of Technology R&D and Projections to 2017. U.S. Department of Energy under Contract DE-AC05-76RL01830 215; 1–25.

[22] Liu WJ, Yu HQ. Thermochemical conversion of lignocellulosic biomass into mass- producible fuels: Emerging technology progress and environmental sustainability evaluation. ACS Environmental Au. 2022;2:98–114.
[23] Awogbemi O, Kallon DVV. Application of biochar derived from crops residues for biofuel production. Fuel Communications. 2023;15(100088):1–22.
[24] AllwynSundarraj AA. Ranganathan TV.A review on cellulose and its utilization from agro-industrial waste. Drug Invention Today. 2018;10(1):89–94.
[25] Mohamed OA. Production of bio-oil from agriculture waste. Biochemistry Analytical Biochemistry. 2018;7(1):1–9.
[26] Pant R, Gupta A, Pant G, Chaubey KK, Kumar G, Patrick N. Second-generation biofuels: Facts and future. Relationship between Microbes and the Environment for Sustainable Ecosystem Services. 2023;3:97–115.

Anil Dhanola

10 Feedstocks for biolubricant production: toward a sustainable future

10.1 Introduction

Population and economic growth are driving a rapid rise in the world's energy consumption. The utilization of petroleum-based products to meet the growing energy demand is being scrutinized due to environmental concerns and the limited availability of fossil resources [1]. There is a push for renewable energy alternatives and energy efficiency in order to reduce reliance on diesel and ensure a sustainable energy future. Lubrication's main goal is to reduce the deterioration of surfaces that are in close contact and move relative to one another [2]. This is accomplished by a substance between the surfaces that functions as a medium to bear the load and keep the surfaces from coming into direct contact with one another. In addition to increasing the life of the surfaces and improving their performance, a substance is used to create a protective coating that lowers friction and enables smooth movement [3]. The name of the substance is "lubricant" and it plays prime role in many industrial applications. Petro-based lubricants and other products made from oil continue to be widely used despite the environmental challenges they cause, which has serious negative effects on the environment. Due to the frequency that these items are used, the environment is being harmed, demanding other methods to mitigate the adverse effects. Additionally, when crude oil sources are depleted, prices rise, forcing the development of substitute goods. This is essential to, among other things, improve sustainability and reduce economic dependency on oil. It becomes crucial to find feasible options for long-term economic and environmental stability. As a result, there is rising concern over the use of such items on a worldwide scale, which has led a few nations and international organizations to push for regulations or agreements that promote the use of ecologically friendly alternatives. This movement for change represents the understanding of the necessity of moving away from items that are destructive to the environment, and toward more sustainable alternatives. Vegetable oil-based biolubricants are becoming more important in a variety of applications. These bio-based lubricants have benefits, such as minimal ecotoxicity, good lubrication properties, high viscosity index (VI), and biodegradability [2, 4]. They are a desirable option for many businesses because of their environmental friendliness. However, bio-based lubricants face obstacles in their large and widespread commercial acceptance in sectors due to their low-temperature characteristics and oxidation stability. One approach to addressing this problem is through chemical modifications of vegetable oils, specifically by eliminating the β-hydrogen atoms in glycerol [5]. The tribological properties of vegetable oils are influenced by factors such as the carbon chain length, the type of fatty acids (FAs) present, and their polarity. By understanding and manipulating these factors, it is possible to enhance the performance of vegetable oil-based lubri-

https://doi.org/10.1515/9783111145136-010

cants and overcome the challenges related to their low-temperature properties and oxidation stability. This chapter serves as a thorough reference, providing essential understanding into the wide range of feedstocks and synthesis procedures involved in the production of biolubricants. It explores the various raw materials employed, highlighting their appropriateness for lubricant production, including vegetable oils, animal fats, and other renewable sources. The physicochemical characteristics of these well-known feedstocks, including their viscosity, oxidative stability, pour point, and other critical features, are also thoroughly covered in this chapter. Researchers and industry experts may make knowledgeable decisions on the choice and use of biolubricants, owing to this knowledge, which is essential for understanding the composition and performance of biolubricants.

10.2 Background and importance of vegetable oil as biolubricants

Lubricants play a vital role in nearly every aspect of modern machinery. Their purpose, as suggested by the name, is to aid the movement of interconnected parts by applying a substance that reduces friction and minimizes wear. The applications of lubricants are diverse: firstly, they diminish wear and heat generated when surfaces are in motion, effectively lowering the friction coefficient between these surfaces. Secondly, they thwart rust and decrease oxidation. Thirdly, they serve as insulators in transformer setups. Lastly, they function as barriers, preventing the ingress of dirt, dust, and water. A lubricant can be defined as a material that diminishes friction and wear by introducing a protective layer between two surfaces in motion. This protective film is what constitutes lubrication when it separates the two surfaces. Lubricants are found in liquid, solid, and gaseous states. An effective lubricant possesses certain characteristics, such as a high melting point, low melting point, high VI, excellent resistance to oxidation, heat endurance, and the ability to guard against corrosion [6]. In real-world scenarios, lubricants can be released into the environment through various means such as leaks, pipe bursts, and unintentional human mistakes. Research indicates that over half of the global lubricant usage is linked to environmental pollution, resulting from practices like total-loss lubrication, spills, and evaporation [7]. Presently, the global production of lubricants stands at 41 million metric tons (MMT). Projections indicate a yearly growth rate of 2% for the market. Notably, the Asia-Pacific region emerges as the leading consumer of lubricants, accounting for approximately 14 MMT annually. Within this, China contributes for 4 MMT and India contributes for 2.4 MMT from the Asian continent, with a substantial portion being lost to the environment [8]. Consequently, there is a strong imperative for lubricants to be both biodegradable and sourced from renewable and sustainable origins.

The term "biolubricant" refers to lubricants that originate from living materials, specifically plant oils and animal fats. The projected growth of the vegetable oil market is outlined, with an anticipated increase from USD 358.14 billion in 2023 to USD 513.44 billion by 2028, reflecting a CAGR of 7.47% for the forecast period of 2023–2028. The surge in demand for vegetable oils is currently driven by their utilization in diverse applications, ranging from fuels and dishwashers to perfume compositions. This surge in demand has been identified as a key catalyst for the market's expansion. For instance, India's utilization of used cooking oil for biolubricants/biodiesel production illustrates this trend, with the volume increasing from 50,000 to 55,000 metric tons between 2020 and 2021, as reported by the USDA Foreign Agricultural Service. Furthermore, the substitution of mineral oil with vegetable oils across various industries as a more environmentally friendly alternative has significantly contributed to the increasing demand. This is evidenced by the commercial availability of vegetable oils for transformers, offering advantages like reduced cost, lower volatility, enhanced pressure performance, higher viscosity, and widespread availability, compared to mineral oils. Leading industry players are capitalizing on this trend by introducing minimally processed and organic variants of vegetable oils. This strategy is bolstering the market's growth trajectory. Additionally, the governments of several nations have introduced initiatives to promote bio-based products like vegetable oils, further fueling this expansion. The molecular structures of all vegetable oils exhibit a striking similarity, characterized by the presence of triglyceride molecules. These triglyceride molecules consist of two fundamental components: the glycerol segment and the FA segment [9]. The glycerol portion is composed of three interconnected hydroxyl groups, while the FA portion consists of long-chain saturated FAs attached to each hydroxyl group. Within vegetable oils, three distinct types of natural FAs can be identified: saturated, monounsaturated, and polyunsaturated FAs as shown in Fig. 10.1. The extent of saturation within vegetable oils hinges on the predominant FAs present in

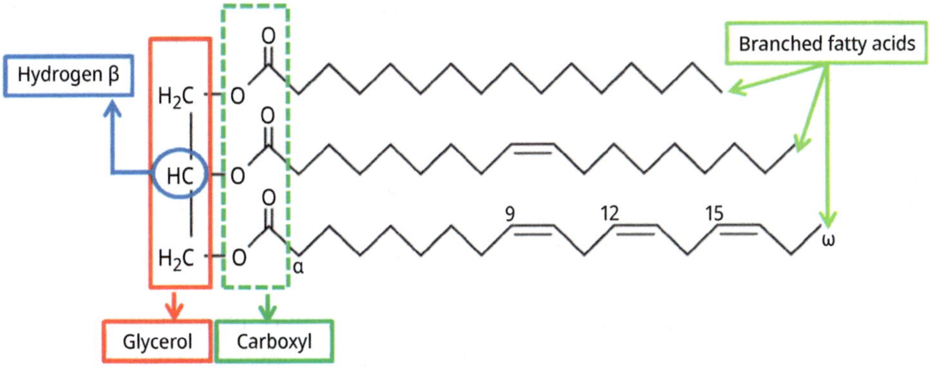

Fig. 10.1: A triglyceride molecule (reused with permission from [5], @ 2021 Elsevier Ltd.).

their composition. This saturation level plays a pivotal role in determining the properties and characteristics of these oils [9].

Vegetable oils can be broadly categorized into two groups: edible and inedible oils. In the contemporary context, the preference has shifted toward utilizing inedible oils to produce bio-based lubricants [5]. This shift is motivated by the fact that inedible oils are sourced from agricultural waste crops, as outlined in reference [10]. This practice ostensibly addresses the concern of diverting food crops for lubricant manufacturing. However, the feasibility of using inedible oils for bio-based lubricants hinges on the availability of ample land to accommodate the cultivation of both edible and inedible crops. Additionally, economic considerations may play a pivotal role in driving farmers to transition from food crop cultivation to the production of biofuel crops, even if the latter is inedible. This unfolding scenario has the potential to create an imbalanced market dynamic. As the production of food crops decreases due to this shift, there is a plausible risk of food prices surging [11]. This underscores the complex interplay between agricultural practices, biofuel production, and food security, warranting a critical examination of the broader implications [5].

The demand for vegetable oil-derived lubricants has intensified across various environmentally conscious industrial practices [12]. Nonetheless, the susceptibility of vegetable oil to oxidation poses a significant challenge to its practicality as a lubricant [12]. The composition of vegetable oils encompasses triglycerides, glycerol molecules, unsaturated FAs, and saturated acids [13]. Triglycerides present in vegetable oil exhibit commendable performance in boundary lubrication scenarios. However, the extent of unsaturation within vegetable oil imposes limitations on its viability and suitability as a lubricant akin to those derived from mineral oils [14]. The stability of a lubricant against oxidation holds paramount importance in its overall lubrication efficacy. Enhancing the inherent properties of a lubricant to align with specific operational requirements and environmental conditions often entails the incorporation of additives. The formulation of a biolubricant hinges on the availability of suitable base stocks. One viable approach to generate a biolubricant involves the esterification of FAs derived from vegetable oil or seed oil, a method elucidated by [15]. This strategy offers a potential solution to overcome the oxidation-related challenges associated with vegetable oil-based lubricants.

10.3 Active vegetable oil feedstocks for biolubricant production

Vegetable oils stand as the primary raw material to produce biolubricants. Thus, it becomes imperative to delve into the chemical structure and characteristics of this foundational resource. The realm of potential feedstocks for biolubricant creation is expansive, with approximately 350 different crops holding the capacity to contribute, due to their oil-rich nature [16]. Among these, there exists a range of edible oils like sunflower oil, rapeseed oil, canola oil, soybean oil, palm oil, and coconut oil [5]. Additionally, inedible vegetable oils such as neem oil, Jatropha oil, rice bran oil, Jojoba oil, mahua oil, and Karanja oil emerge as viable candidates to serve as foundational components for the formulation of biolubricants. This diversity in feedstock sources opens avenues for the development of sustainable and effective biolubricant solutions. Historically, vegetable oils have fulfilled the role of lubricants and hold a distinct edge over their mineral-based counterparts due to their biodegradability and nontoxic nature. In recent times, a heightened environmental awareness within industries and the general populace has emerged. This has led to the practical adoption of vegetable-based lubricants across various applications. The advantages of vegetable-based oils over petroleum- or mineral-based oils are pronounced: they exhibit a higher VI, reduced evaporation loss, and the potential to enhance lubricity, ultimately contributing to enhanced energy efficiency. Embracing vegetable oils as a lubricant choice is an evident decision, attributed to their impressive biodegradability and nontoxic properties. Furthermore, their extraction from renewable sources reinforces their attractiveness. Diverse physicochemical attributes of vegetable oils and fats result in a spectrum of applications, often in competition with petroleum derivatives, contingent on carbon chain distribution. In the realm of seed oils, a consistent set of the five most vital FAs – C16:0 (palmitic), C18:0 (stearic), C18:1 (oleic), C18:2Δ9,12 (linoleic), and C18:3Δ9,12,15 (linolenic) is present in the structural lipids of plant tissues [17]. Some seed oils encompass monoesters of long-chain FAs and fatty alcohols that exhibit varying chain lengths and degrees of unsaturation. In the contemporary landscape of lubricant feedstocks, rapeseed (canola) oil, soybean oil, and sunflower oil have emerged as globally dominant choices for production. Rapeseed oil (sourced primarily from Europe, China, and Canada) ranks as the third most significant vegetable oil source worldwide.

10.3.1 Palm oil

Malaysia stands as a prominent figure in the production and global trade of palm oil. As one of the world's leading palm oil producers and exporters, Malaysia significantly bolsters the global palm oil supply. The unique composition of palm oil, rich in FAs, adds to its allure. Palm oil's utility as a lubricant can be context-dependent. Palm oil containing high levels of saturated FAs (SFAs) tends to exhibit less desirable flow characteristics at lower temperatures (around 8–15 °C), limiting its use in lubrication, especially in colder regions. However, in tropical climates, the same high SFA content can prove advantageous due to its positive impact on oxidative stability. Further innovations are being explored, such as the development of high-oleic palm trimethylolpropane esters. These esters offer promising potential as base stocks for biodegradable lubricants. They possess suitable pour points and lubrication properties, making them an intriguing candidate for applications in the field of environmentally friendly lubricants [18].

10.3.2 Coconut oil

Coconut oil presents itself as clear and pale-yellow oil. Classified as a member of lauric oils, it boasts substantial quantities of C12:0 (around 50%), and is rich in saturated short-chain FAs (approximately 85–90%). This composition results in coconut oil having a notable pour point of 24 °C. This elevated pour point can be attributed to the prevailing saturation within its FA constituents [19]. In the realm of vegetable oils, coconut oil stands out for its elevated pour point and tendency to exhibit increased wear when employed as a lubricant in metallic interfaces. This characteristic combination has imposed limitations on its industrial applications, curtailing its potential utilization within various sectors [20].

10.3.3 Sunflower oil

Sunflower oil ranks as the fourth most significant vegetable oil in terms of global production, contributing a substantial 9 MMT, which accounts for approximately 8% of the total vegetable oil output. The diverse oleic content within sunflower oil categorizes it into three main types: (1) regular (14–39%), (2) mid-oleic (43–72%), and (3) high-oleic (75–91%) [21]. In exploring its applications, it has been observed that lubricants derived from sunflower oil demonstrate notable efficacy in reducing emission levels of carbon monoxide (CO), carbon dioxide (CO_2), and hydrocarbons across various testing scenarios. This suggests a promising potential for sunflower oil to serve as a lubricant for interacting components, indicating a novel avenue for its utilization [22].

10.3.4 Soybean oil

Soybean oil emerges as a budget-friendly foundational component for a diverse spectrum of lubricating fluids, all the while boasting an environmentally positive impact. Incorporating soybean oil into lubricant formulations is not only economically sensible but also holds compelling performance benefits. Soybean oil exhibits an innate propensity to dissolve polar impurities with greater ease compared to mineral-based fluids. This attribute curtails the necessity for detergent additives, aligning with the broader objective of reducing additives in lubricants. Moreover, soybean oil showcases robust compatibility with various oil additives that contribute to crucial lubricant characteristics such as lubricity, antiwear protection, load-carrying capacity, rust prevention, and control of foaming [23, 24].

10.3.5 Sesame oil

Sesame oil, sourced from sesame seeds, stands as an edible vegetable oil with a historical legacy, being one of the earliest crop-derived oils. It is often referred to as the "queen of oilseed crops," highlighting its significance. What sets sesame oil apart is its relative rarity among readily available vegetable oils. This rarity does not hinder its practicality, as it boasts a low pour point, thanks to its inherent unsaturation. Furthermore, its commendable oxidative stability can be attributed to the presence of natural antioxidants like sesamin, sesamolin, sesamol, sesaminol, and γ-tocopherol [25]. However, while preliminary investigations have been carried out, it is worth noting that initial findings indicate that raw sesame oil presents limitations in terms of its tribological properties, as highlighted in earlier studies [26].

10.3.6 Castor oil

Castor oil boasts a rich heritage when it comes to lubrication practices, spanning back through time. Extracted from the seeds of the castor plant (*Ricinus communis*), through pressing, castor oil takes the form of colorless to exceedingly pale-yellow liquid. Its aroma and taste are mild or virtually imperceptible. A distinguishing feature of castor oil lies in its impressive performance attributes. It excels not only in low-temperature viscosity characteristics but also in high-temperature lubrication properties, surpassing the capabilities of numerous other vegetable oils. These attributes render castor oil an asset in the realm of lubrication. Comparatively, castor oil outperforms most vegetable oils, displaying superior low-temperature viscosity and high-temperature lubrication capabilities, as highlighted by prior research [27].

10.3.7 Jatropha oil

Jatropha, previously not associated with food production, has garnered significant attention in its recent role as a key player in fuel applications, particularly in the realm of Jatropha biodiesel. Among the multitude of oil-yielding species, Jatropha stands out for its remarkable promise, boasting one of the most impressive potentials [28]. Yet, it is worth noting that Jatropha does face a certain limitation, notably its elevated C18:2 content. However, this should not overshadow its merits. Its substantial FA content, ranging from 61% to 64%, positions Jatropha oil as a prime candidate for potential use in lubricant production [29].

10.3.8 Karanja oil

Karanja oil has emerged as a noteworthy contender in the realm of feedstocks, showing significant promise for large-scale biodiesel production. This can be attributed to its remarkable adaptability to prevailing climatic conditions, making it a surplus resource across the expanse of the South Asian region [30]. When evaluated in terms of cost, Karanja oil presents a reasonable alternative in comparison to other oils. Notably, the extraction of Karanja oil employs the cold-press extraction method. The applications of Karanja oil stretch beyond biodiesel. Its chemically modified variations also find utility as both a lubricant and biofuel [31]. However, it is important to acknowledge that the oil's lubricity might not achieve its full potential due to certain limitations arising from its oxidation stability.

10.3.9 Linseed oil

Linseed oil boasts a distinctly favorable FA composition, predominantly due to its substantial linolenic acid content. Notably, linseed oil is rich in α-linolenic acid, constituting nearly 60% of its composition, a noteworthy contrast to the average of 25% found in other plant oils. The observed advantageous attributes of linseed oil are likely attributed to its elevated content of triglycerides featuring linoleic acid. This composition manifests in a relatively high number of unsaturated bonds, which might contribute to its perceived benefits [32]. In essence, linseed oil stands as a predominantly composed blend of various FAs, encompassing both saturated and unsaturated forms. This mixture includes constituents like palmitic acid, stearic acid, oleic acid, and linoleic acid [33].

10.3.10 Rice bran oil

Rice bran oil is derived from the outer husk of rice grains, making it a compelling option as base oil for specific applications. This choice stems from its distinctive FA composition. Notably, rice bran oil stands out due to its relatively lower proportions of linoleic acid and linolenic acid, combined with a higher percentage of oleic acid. This composition contributes to a pronounced reduction in friction [34]. With a significant presence of 41.86% oleic acid, rice bran oil emerges as a promising candidate for the advancement of lubricant base oils. The FA makeup of rice bran showcases a blend of constituents, including 24% SFAs, 41.86% monounsaturated FAs, and 30.99% polyunsaturated FAs [35].

10.3.11 Neem oil

Neem oil, extracted from the seeds of the *Azadirachta indica* tree, has garnered global attention for its versatile applications across diverse regions. Its multifaceted utility spans domains such as medicine and agriculture. This natural resource has been harnessed for an array of purposes, ranging from soap manufacturing to serving as a foundational ingredient in the creation of commercial pesticides and cosmetics. Additionally, neem oil finds significance in areas like plant protection, textile safeguarding, stock preservation, engine lubrication, lamp oil, and even the production of candles [36]. A distinctive feature of neem oil is its color variability. It exhibits a spectrum of hues, spanning from golden yellow to yellowish brown, reddish brown, dark brown, greenish brown, and at times, even a vibrant red. This diversity in color mirrors the extensive range of applications and benefits that neem oil offers.

10.4 Properties

The effectiveness and operational efficiency of lubricants are intricately linked to the distinctive physical and chemical attributes they possess. The way these lubricants interact with surfaces and the caliber of lubrication they provide are predominantly shaped by these specific properties. Figure 10.2 presents a comprehensive array of

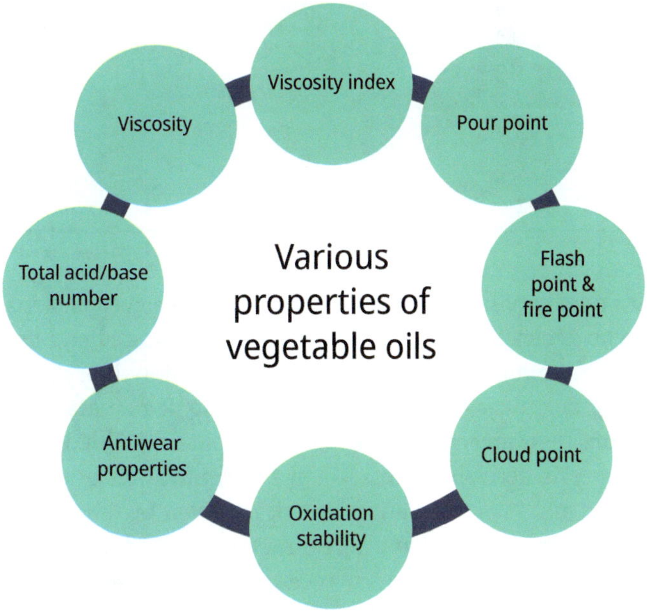

Fig. 10.2: Comprehensive array of properties.

these properties, each of which plays a distinct role in defining the behavior of a given lubricant. These properties, collectively harmonizing within a single lubricant, together dictate its capacity to engage with surfaces effectively, foster optimal lubrication, and consequently, contribute to the overall performance and longevity of the mechanical systems involved. Extensive investigations have delved into how the physicochemical properties of vegetable oils are influenced by their unique compositions and structures. A diverse range of vegetable oils has undergone thorough assessment, primarily focusing on their behavior concerning thermal dynamics, oxidation susceptibility, and performance at low temperatures. The objective behind such scrutiny is to ascertain their viability as fundamental fluids for lubrication applications in both industrial and automotive contexts. This exploration places considerable emphasis on comprehending the intricate interplay between structure, properties, and functionality. This understanding serves as a pivotal factor in crafting an optimal oil composition that aligns seamlessly with specific applications, tailoring its attributes to best suit the intended purpose. Table 10.1 shows the important properties of various vegetable oils.

Tab. 10.1: Important properties of various active vegetable oils [37].

Vegetable oil	Viscosity at 40 °C (cSt)	Viscosity index	Pour point (°C)	Flash point (°C)	Oxidation stability (min)
Soybean oil	28.86	246	−9	325	–
Sunflower oil	40.05	206	−12	252	–
Sesame oil	27.33	193	−5	316	–
Rice bran oil	40.6	169	−13	318	–
Rapeseed oil	45.60	180	−12	252	–
Palm oil	52.4	186	−5	–	–
Moringa oil	44.9	–	–	204	28.27
Jatropha oil	35.4	205	−6	186	5
Coconut oil	24.8	169	21	325	–
Castor oil	220.6	220	−27	250	–
Cottonseed oil	33.86	211	–	252	–

Until now, the predominant endeavors in the realm of vegetable oils have centered on comprehending the physicochemical attributes intrinsic to both saturated and unsaturated FAs, primarily with the intent of utilizing them as eco-friendly lubricants. These oils offer the versatility of direct usage or incorporation as supplementary components in specific formulations, contingent upon the specific context [38]. Concurrently, the scientific community is actively engrossed in the innovation of novel additives to blend into vegetable oils. This innovative pursuit is driven by the aspiration to augment a spectrum of their physico-chemical characteristics, encompassing factors such as thermo-oxidative stability and performance in sliding contact scenarios, among various other considerations [39].

10.5 Chemical modification methods of vegetable oils

Until now, most studies on vegetable oil-based biolubricants have concentrated on enhancing their physicochemical and lubrication properties using various chemical synthesis/diverse catalytic techniques. Nevertheless, it is crucial to persist in the advancement of these biolubricants, as they still show low-temperature fluidity and chemical stability due to the unsaturated FA structure [40]. The most encouraging approaches revolve around altering unsaturated oils through chemical processes, as the double bonds within them are susceptible to reactions with atmospheric oxygen. The various synthesis techniques for modifying vegetable oils encompassing various methods along with their merits and demerits are shown in Fig. 10.3.

10.5.1 Esterification/transesterification

The production of biolubricants from oils involves two processes; esterification and transesterification. These processes are essential for converting oils into biolubricants with improved properties. Esterification is a reaction where an alcohol combines with an acid (or a derivative of an organic acid) to form an ester and water. In the context of production, this process involves reacting a vegetable oil (which contains FAs) with an alcohol (methanol or ethanol) using a catalyst. This results in the formation of FA alkyl esters, commonly known as biodiesel, which can be used as biolubricant due to its lubricating properties. On the hand, transesterification refers to the exchange of one ester group in an ester molecule with another alcohol molecule. In the realm of biolubricants, transesterification usually refers to the reaction between a vegetable oil and an alcohol, methanol, or ethanol. This reaction breaks down the triglycerides in the oil into FA alkyl esters and glycerol. These FA alkyl esters, similar to those produced through esterification, serve as the components of biolubricants. Typically, the catalysis of esterification and transesterification reactions involves mineral acids like HCl, H_2SO_4, or H_3PO_4 [37]. These reactions can take place utilizing either acidic or basic catalysts. Although transesterification has been thoroughly examined in laboratory settings with an emphasis on acidic conditions, its adoption on a commercial scale remains infrequent. Both esterification and transesterification processes are commonly employed in the production of biolubricants from vegetable oils due to their effectiveness in reducing viscosity and enhancing lubricating properties. These processes not only contribute to the improvement of the overall quality of biolubricants but also promote the use of renewable resources, making them more environmentally friendly alternatives to conventional lubricants derived from fossil fuels.

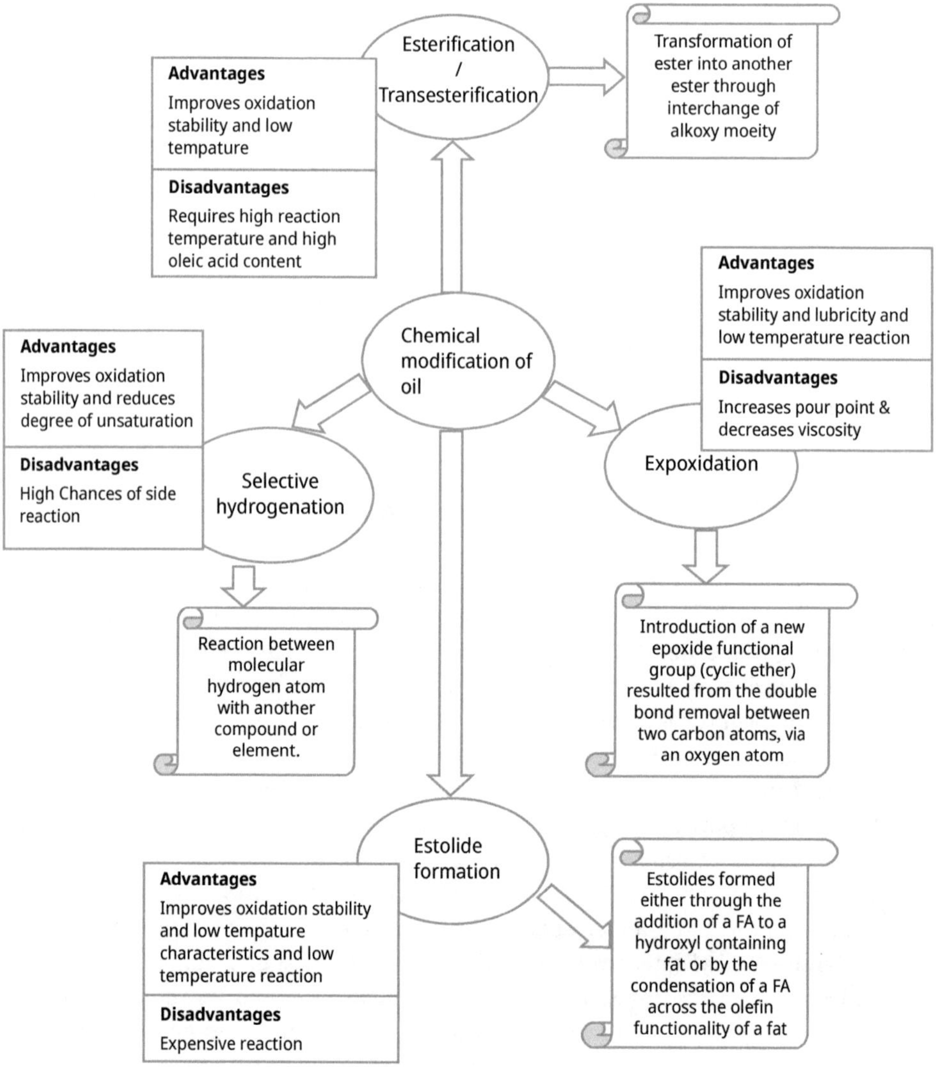

Fig. 10.3: Various chemical synthesis techniques of vegetable oils [41].

10.5.2 Epoxidation

Epoxidation stands as a critical procedure in the modification of unsaturation, aiming to produce biolubricants, boasting remarkable qualities such as a superior pour point, heightened oxidative stability, enhanced metal surface adsorption, elevated viscosity, improved lubricity, and a notable VI. Epoxidized vegetable oils come into existence through a chemical reaction involving the utilization of peroxy acids to target and transform the double bonds [10], ultimately resulting in the removal of these C = C bonds [42]. Considering the inherent temperature-related limitations of epoxidized products, there is a need to enhance their physicochemical attributes. This can be achieved by subjecting these epoxidized compounds to modifications through processes such as esterification reactions, as well as techniques like alkylation, acylation, acyloxylation, amino alkylation, cooligomerization, enereaction, hydroaminomethylation, or hydroformylation [43, 44].

10.5.3 Estolides

The estolide process is a method used in the production of biolubricants. It involves creating estolides, which are bio-based esters made from FAs and their derivatives. These estolides have lubricating properties, stability against oxidation, and viscosity characteristics, making them excellent choice for applications. To start the estolide process, FAs are reacted with FA derivatives like FA anhydrides or esters under controlled conditions. This reaction leads to the formation of estolide molecules through processes like esterification and transesterification. The resulting estolides have a structure that contributes to their enhanced properties, when compared to traditional linear esters. In this context, the carbocation mentioned becomes open to interactions with nucleophiles, triggered by FAs. As a result, an ester compound called "estolide" is formed. In a sense, estolide compounds are known for their lubricating properties, increased resistance to heat damage, better VI, and lower pour point in the resulting biolubricant [45].

10.5.4 Hydrogenation

Hydrogenation is a process that involves adding hydrogen atoms to the composition of oil. This procedure is carried out with the help of catalysts, which aim to improve the resistance of oil, to oxidation. Through hydrogenation, linoleic acid transforms into cis-structured acid. Then, oleic acid becomes a generic FA. As a result, there is an increase in saturated FAs and a decrease in FAs in the composition of the vegetable oil [46].

10.6 Conclusion

In the present circumstances, there exists a significant demand for the advancement of biolubricants that are not only biodegradable but also eco-friendly. Globally, researchers are significantly prioritizing the exploration and advancement of biolubricants that hold broad-ranging potential for use across diverse industrial domains. These biolubricants not only offer versatility but also have the potential to serve as viable alternatives to conventional mineral-based lubricants. Vegetable oils have emerged as ideal contenders for biolubricant production, owing to their exceptional lubricating properties. Transesterification, esterification, estolide formation, epoxidation, hydrogenation, and similar processes constitute the primary methodologies employed in biolubricant production. The resultant biolubricants are influenced by factors such as the catalyst used, reaction temperature, and duration of the process. Numerous research endeavors have delved into the properties of diverse vegetable oils. Nonetheless, the pursuit of optimal and high-performing lubricants tailored to specific mechanical components remains a formidable challenge within the current landscape of the lubricant industry. The necessity for next-generation lubricants within industries has grown significantly, driven by the increasing demand for effectively lubricated mechanical systems.

References

[1] Zulqarnain Mohd Yusoff MH, Ayoub M, Ramzan N, Nazir MH, Zahid I, Abbas N, Elboughdiri N, Mirza CR, Butt TA. Overview of feedstocks for sustainable biodiesel production and implementation of the biodiesel program in Pakistan. ACS Omega. 2021;6(29):19099–114.

[2] Dhanola A, Garg HC. Influence of different surfactants on the stability and varying concentrations of TiO_2 nanoparticles on the rheological properties of canola oil-based nanolubricants. Applied Nanoscience. 2020;10:3617–37.

[3] Dhanola A, Khanna N, Gajrani KK. A critical review on liquid superlubricative technology for attaining ultra-low friction. Renewable and Sustainable Energy Reviews. 2022;165:112626.

[4] Mobarak HM, Mohamad EN, Masjuki HH, Kalam MA, Al Mahmud KA, Habibullah M, Ashraful AM. The prospects of biolubricants as alternatives in automotive applications. Renewable and Sustainable Energy Reviews. 2014;33:34–43.

[5] Zainal NA, Zulkifli NWM, Gulzar M, Masjuki HH. A review on the chemistry, production, and technological potential of bio-based lubricants. Renewable and Sustainable Energy Reviews. 2018;82:80–102.

[6] Saboya RM, Cecilia JA, García-Sancho C, Sales AV, de Luna FM, Rodríguez-Castellón E, Cavalcante CL Jr. Assessment of commercial resins in the biolubricants production from free fatty acids of castor oil. Catalysis Today. 2017;279:274–85.

[7] Rudnick LR, ed. Synthetics, Mineral Oils, and Bio-based Lubricants: Chemistry and Technology. Laramie, USA: CRC press, 202029.

[8] https://www.mosil.com/blog/wp-content/uploads/2020/01/Development-of-Lubricant-Industry-in-next-5-years.pdf

[9] Syahir AZ, Zulkifli NW, Masjuki HH, Kalam MA, Alabdulkarem A, Gulzar M, Khuong LS, Harith MH. A review on bio-based lubricants and their applications. Journal of Cleaner Production. 2017;168:997–1016.

[10] Atabani AE, Silitonga AS, Ong HC, Mahlia TM, Masjuki HH, Badruddin IA, Fayaz H. Non-edible vegetable oils: A critical evaluation of oil extraction, fatty acid compositions, biodiesel production, characteristics, engine performance and emissions production. Renewable and Sustainable Energy Reviews. 2013;18:211–45.

[11] Kenney VP, Erichsen RL. Conflict between fuel and food: The ethical dimension. Energy in Agriculture. 1983;2:285–306.

[12] Delgado MA, García-Rico C, Franco JM. The use of rosemary extracts in vegetable oil-based lubricants. Industrial Crops and Products. 2014;62:474–80.

[13] Syahrullail S, Kamitani S, Shakirin AJ. Performance of vegetable oil as lubricant in extreme pressure conditions. Procedia Engineering. 2013;68:172–77.

[14] Somidi AK, Das U, Dalai AK. One-pot synthesis of canola oil-based biolubricants catalyzed by MoO3/Al2O3 and process optimization study. Chemical Engineering Journal. 2016;293:259–72.

[15] McNutt J. Development of biolubricants from vegetable oils via chemical modification. Journal of Industrial and Engineering Chemistry. 2016;36:1–2.

[16] Kurre SK, Yadav J. A review on bio-based feedstock, synthesis, and chemical modification to enhance tribological properties of biolubricants. Industrial Crops and Products. 2023;193:116122.

[17] Bart JC, Gucciardi E, Cavallaro S. Renewable feedstocks for lubricant production. Biolubricants: Science and Technology. 2013;121–248. doi: https://doi.org/10.1533/9780857096326.121.

[18] Yunus R, Fakhru'l-Razi A, Ooi TL, Iyuke SE, Perez JM. Lubrication properties of trimethylolpropane esters based on palm oil and palm kernel oils. European Journal of Lipid Science and Technology. 2004;106(1):52–60.

[19] Ajithkumar G, Jayadas NH, Bhasi M. Analysis of the pour point of coconut oil as a lubricant base stock using differential scanning calorimetry. Lubrication Science. 2009;21(1):13–26.

[20] Sajeeb A, Rajendrakumar PK. Comparative evaluation of lubricant properties of biodegradable blend of coconut and mustard oil. Journal of Cleaner Production. 2019;240:118255.

[21] Grompone MA Sunflower Oil. Vegetable oils in food technology: Composition, properties and uses. Gunstone, FD, Ed.

[22] Jabal MH, Abdulmunem AR, Abd HS. Experimental investigation of tribological characteristics and emissions with nonedible sunflower oil as a biolubricant. Journal of the Air & Waste Management Association. 2019;69(1):109–18.

[23] Schmitz JF, Erhan SZ, Sharma BK, Johnson LA, Myers DJ. Bio-based products from soybeans. In: Soybeans. AOCS Press, 2008, 539–612.

[24] Srivastava A, Sahai P. Vegetable oils as lube base stocks: A review. African Journal of Biotechnology. 2013;12(9):880–891.

[25] Fukuda Y, Osawa T, Kawakishi S, Namiki M. Chemistry of lignan antioxidants in sesame seed and oil. ChemInform. 1994;25(15):264–274.

[26] Nair SS, Nair KP, Rajendrakumar PK. Evaluation of physicochemical, thermal and tribological properties of sesame oil (Sesamum indicum L.): A potential agricultural crop base stock for eco-friendly industrial lubricants. International Journal of Agricultural Resources, Governance and Ecology. 2017;13(1):77–90.

[27] Ogunniyi DS. Castor oil: A vital industrial raw material. Bioresource Technology. 2006;97(9):1086–91.

[28] Bart JC, Palmeri N, Cavallaro S. Biodiesel Science and Technology: From Soil to Oil. Cambridge: Elsevier, 201019.

[29] Akbar E, Yaakob Z, Kamarudin SK, Ismail M, Salimon J. Characteristic and composition of Jatropha curcas oil seed from Malaysia and its potential as biodiesel feedstock. European Journal of Scientific Research. 2009;29(3):396–403.

[30] Agarwal AK, Dhar A. Karanja oil utilization in a direct-injection engine by preheating. Part 1: Experimental investigations of engine performance, emissions, and combustion characteristics. Proceedings of the Institution of Mechanical Engineers, Part D: Journal of Automobile Engineering. 2010;224(1):73–84.

[31] Dhar A, Agarwal AK. Experimental investigations of effect of Karanja biodiesel on tribological properties of lubricating oil in a compression ignition engine. Fuel. 2014;130:112–19.

[32] Myshkin NK, Grigoriev AY, Kavaliova IN. Influence of composition of plant oils on their tribological properties. Tribology in Industry. 2017;39(2):207.
[33] Gruia A, Raba DN, Dumbrava D, Moldovan C, Bordean D, Mateescu C. Fatty acids composition and oil characteristics of linseed (Linum Usitatissimum L.) from Romania. Journal of Agroalimentary Processes and Technologies. 2012;18(2):136–40.
[34] Sneha E, Akhil RB, Krishna A, Rani S, Kumar SA. Formulation of bio-lubricant based on modified rice bran oil with stearic acid as an anti-wear additive. Proceedings of the Institution of Mechanical Engineers, Part J: Journal of Engineering Tribology. 2021;235(9):1950–57.
[35] Rani S, Joy ML, Nair KP. Evaluation of physiochemical and tribological properties of rice bran oil–biodegradable and potential base stoke for industrial lubricants. Industrial Crops and Products. 2015;65:328–33.
[36] Ozioko FU. Extraction and characterization of soybean oil based bio-lubricant. AU Journal of Technology. 2012;15(4):260–264.
[37] Cecilia JA, Ballesteros Plata D, Alves Saboya RM, Tavares de Luna FM, Cavalcante CL Jr, Rodríguez-Castellón E. An overview of the biolubricant production process: Challenges and future perspectives. Processes. 2020;8(3):257.
[38] Fox NJ, Tyrer B, Stachowiak GW. Boundary lubrication performance of free fatty acids in sunflower oil. Tribology Letters. 2004;16(4):275–81.
[39] Choi US, Ahn BG, Kwon OK, Chun YJ. Tribological behavior of some anti-wear additives in vegetable oils. Tribology International. 1997;30(9):677–83.
[40] Birova A, Pavlovičová A, Cvenroš J. Lubricating oils based on chemically modified vegetable oils. Journal of Synthetic Lubrication. 2002;18(4):291–99.
[41] Joshi JR, Bhanderi KK, Patel JV. A review on bio-lubricants from non-edible oils-recent advances, chemical modifications and applications. Journal of the Indian Chemical Society. 2022;14:100849.
[42] Tan SG, Chow WS. Bio-based epoxidized vegetable oils and its greener epoxy blends: A review. Polymer-Plastics Technology and Engineering. 2010;49(15):1581–90.
[43] Salih N, Salimon J. A review on new trends, challenges and prospects of ecofriendly friendly green food-grade biolubricants. Biointerface Research in Applied Chemistry. 2021;12(1):1185–207.
[44] Karmakar G, Ghosh P, Sharma BK. Chemically modifying vegetable oils to prepare green lubricants. Lubricants. 2017;5(4):44.
[45] Cermak SC, Bredsguard JW, John BL, McCalvin JS, Thompson T, Isbell KN, Feken KA, Isbell TA, Murray RE. Synthesis and physical properties of new estolide esters. Industrial Crops and Products. 2013;46:386–91.
[46] Mathew BC, Thangaraja J. Material compatibility of fatty acid methyl esters on fuel supply system of CI engines. Materials Today: Proceedings. 2018;5(5):11678–85.

Chandra Mohan, Jenifer Robinson

11 Life cycle analysis of bio-based lubricants

11.1 Introduction

A peripheral portion or a space of an object that can be observed or interacted with any other material is called the surface. These surfaces can be generally classified as smooth and rough. All objects experience the frictional force between any two surfaces in contact. Through simple experiments, it has been proved that smooth surfaces experience less frictional forces than rough surfaces. Abnormalities between surfaces cause variations in frictional forces that are in interaction and repel the motion between the two objects. The characteristics of lubricants are to reduce friction and adhesion, thereby acting as glidants, coolants, shock absorbers, cleaning agents, sealants, cushioning agents, and antioxidants. Lubricants enhance the additional lifespan of all machines and heavy equipment. These lubes help improve the performance of machinery and are very cost-effective. When two rough surfaces are in motion, a large amount of heat is generated. Lubricants, when applied, form a thin film between the surfaces and hence can reduce the heat generated. Mostly oils such as engine oil, hydraulic oil, and greases act as good lubricants [1, 2]. The main factor is the application of lubricants, which should be done with utmost care. Multilayers of lubricants between any two sliding surfaces may reverse the mechanism and increase the frictional heat. To prevent this state, unimolecular lubricant films are recommended to reduce friction to a greater extent. By decreasing the heat generated, irregularities present between two heavy surfaces are removed completely in industrial machines. The choice of a suitable lubricant is another key factor in any industry. Many hydraulic-based fluids are used in vehicles that can transform force into pressure and magnify the decelerating forces. The base oils with additives are used as improvers for internal ignition engines, neutralizing acids released from fuels. These oils act as heat absorbers, cooling the piston loops and moving parts of the engine [3].

11.2 Bio-based lubricants

Commonly known as "biolubes," they are partly derived from biological materials and blended with synthetic base oils. There are three types of bio-based lubricants: (i) plant-based, (ii) animal-based, and (iii) sustainable forestry-based (Tab. 11.1).

It has been found that there are three important benefits of using biolubricants in industries – reduction of friction and wear, reduction of operating noise, and reduction of heat generated, finally leading to enhanced industrial appliances. According to the World Market report, the three major oils used globally are engine oils

https://doi.org/10.1515/9783111145136-011

Tab. 11.1: Some biolubricants: types and applications.

S. no.	Lubricating oils	Types	Applications
1	Jojoba oil	Plant-based	Lubricates machine guns
2	Beef tallow oil	Animal-based	Jet fuel
3	Soy oil	Forestry-based	Hydraulic fluid to the elevator system
4	Bio-oil	Semisynthetic-based	Fuels in boilers, engines, and turbines
5	Gear oil	Semisynthetic-based	Transmitters, manual gearboxes, and vehicle gear systems
6	Biodiesel	Semisynthetic-based [4]	Transportation fuel in power engines, fuel filters, and heating oils [5]
7	Biogrease	Semisynthetic-based [6]	Green lubricants and marine sectors
8	Bio-steering fluid	Semisynthetic-based	Power steering systems

(50%), process oils (15%), and hydraulic oils (10%); other remaining oils make up 25%. Biodegradable lubricants are generally manufactured with high corrosion resistance and high oxidation stability. The versatile applications of biolubricants are found in industries involving food processing, industrial engineering, farming, automotive, wastewater, aerospace, commercial, oceanic, and defense. The high demand for bio-based lubricants is increasing due to adoption of new green technologies, intense funding of research work, reduction in usage of petroleum imports, and sources for a healthy environment.

Biodegradable lubricants are naturally vegetable-oil based. Novel processing techniques are emerging that mainly focus on green approaches and sustainable green chemical modifications. As synthetic lubricants are characteristically toxic and are non-eco-friendly in nature, firm government regulations are being implemented toward making pollution-free lubricants. High viscosity index (VI), low volatility, good lubricity, high oxidation stability, high thermal stability, and high corrosion resistance are the preferred properties during the commercial processing of vegetable oil lubricants. However, the presence of double and triple bonds has lowered these properties and natural vegetable oils are tested with poor oxidation and thermal stabilities. Several suggested chemical modifications have been attempted through transesterification, selective hydrogenation, and epoxidation. Among these chemical methods, epoxidation of natural vegetable oils is used in the processing of bio-based lubricants. The advantages of the epoxidation method are optimum reaction conditions, formation of stable desired products, stereoselective nature, minimum side reactions and by-products, and feasible conversions. The synthesis of bio-based lubricants from epoxidized vegetable oil using suitable alcohols and specific catalysts has resulted in the manufacturing of nontoxic and stable biolubes, which are proven through the viscosity measurements and pour point tests [7]. Alternatively, researchers are planning for hereditarily improved natural vegetable oilseeds that have high oxidation, viscosity, and thermal stability with cost-effective approaches, so that the addition of synthetic additives and other chemical modifications can be entirely eliminated in the making of bio-based lubricants. Based on the regulations of the Food and Drug Administration (FDA), safe biolubes both for individuals and types of machinery are expected soon.

Eco-lube greases are generally made from high-quality oilseeds grown in superior conditions followed by treatment with antioxidants, anti-wear, and anticorrosive additives. The noticeable advantages are enhanced friction reduction, enhanced load capacity, and reduction in wear and tear and corrosion of machinery. These greases are basically problem-solving and can be used in any working conditions related to portable water contact, incidental food contact, and higher temperatures. These are eco-friendly biodegradable formula-applied greases that are ideal for protecting all types of machinery made of aluminum, galvanized and tinned steel, stainless steel, and chrome-plated machines, and can be cleaned off easily using any soap and water.

11.3 Life cycle assessment of biolubricants

Life cycle assessment (LCA) is a set of procedures for assessing the physical and chemical properties of a new product that contributes to the environmental factors in its lifetime, thereby increasing resource-use effectiveness and decreasing accountabilities. The five major steps in LCA are: (i) analysis of the new product, (ii) production progression, (iii) applications, (iv) cessation period, and (v) economy aspects. LCA is done to ensure the environmental effects of manufacturing a new product based on cost-effectiveness, eco-friendliness, or performance. The framework of LCA will support every company or industry to acquire complete analysis to assess techniques related to various stages of a new product's life. During LCA, setting up the goal and scope of the new product, inventory analysis, the impact of the assessment and its analysis, interpretation of the assessment, further product development and improvement methods, planning and making new supporting/ governmental policies, marketing strategies, any other related issues and challenges resulting in appropriate solutions, and finally its applications in various fields are scrutinized thoroughly. LCA aids companies and industries in making the right decisions on the environmental impacts of the expected product, raw materials to be used, design, thinking, and formulating concrete policies. To understand the strengths and weaknesses of the desired product, the LCA procedures are generally classified as: (i) abstract LCA, (ii) basic LCA, and (iii) broad LCA.

LCAs of bio-based lubricants and mineral-based lubricants were performed particularly on the environmental impact characteristics. Mineral-based lubricants cause severe environmental destruction damaging waterbodies, releasing highly toxic substances, and are nonbiodegradable. According to LCA reports, bio-based lubricants that are obtained from plant oilseeds prevent eutrophication, air-, and water impacts, use minimal additives, control acidification, reduction in photochemical smog potential, and lower contribution to global warming and ozone depletion. Biolubricants are renewable and biodegradable and so researchers are working on the improvement of their properties and production processes to enhance their applications. The life cycle studies on the new chemical modification pathways and the addition of suitable additives to improve sustainability are still under review [8]. The investments in biolubricants are comparatively lower at the native level than globally in terms of environmental and economic points. Research is still in progress on the ecological footprints of conventional and bio-based lubricants to raise awareness among the end-users and for future developments.

A sustainability assessment of biolubricants [9] has been established to study the environmental, social, and economic effects of a new product. The framework focuses on qualitative and quantitative analysis of the product in deeper aspects than with simple data analysis. Initially, a tool is designed, which will be used to test all types of products. Based on the test conducted, the descriptive indicators are framed to assess the process by the particular peripheral shareholders. Thus, the tool proposes rapid decision-making on the new products to be promoted in a sustainable direction.

11.4 Case studies: Life cycle analysis of biolubricants

11.4.1 Life cycle analysis of biolubricants for aluminum rolling

Biolubricants [10] are renewable, nontoxic, biodegradable, and easily mined and administered. To understand the influences of biolubricants on society, the environment, and technology, the properties of biolubricants should be examined, issues and challenges related to volatile emissions and solid wastes are to be measured and compared, and finally the extent of reuse and degree of production to prove that biolubricants are preferred to petro-based lubricants have to be assessed.

The outputs of LCA on bio-based lubricants have proven less volatile emissions, biodegradability, free of aromatics and carcinogenic substances, increased lubricity, machine compatibility, increased stability and volatility, higher VI, and nontoxicity towards the environment (Tab. 11.2).

Tab. 11.2: LCA report: bio-based lubricants versus mineral lubricants.

Various characteristics	Bio-based lubricants	Mineral lubricants
Raw material	Soybean oilseeds, fertilizers, pesticides, lime, nitrogen, carbon dioxide, and energy	Crude oil and energy
Emissions and wastes	Nitrates, oxides of nitrogen, and other combustion by-products	Methane gas and combustion by-products
Extraction processing	Hexane, energy, and steam	Energy and steam
Emissions and wastes	Combustion by-products and volatile organic compounds (VOCs)	Combustion by-products and non-combustion emissions
Aluminum rolling	Soybean oil	Mineral oil
Emissions and wastes	VOCs	VOCs
Properties tested		
– *Acidification*	Low	High
– *Eutrophication*	Low	High
– *Human health*	Nontoxic	Toxic
– *Smog formation*	Low	High
– *Global warming*	Less CO_2 emissions	More CO_2 emissions
– *Use of fossil fuel*	Minimum	Maximum
– *Climate change*	Minimum	Maximum
Use rate	Lower	Higher
Performance	Higher	Lower

11.4.2 Life cycle analysis of industrial biolubricants

In 2019, the world's first industrial biolubricant [11] with nearly 100% bio-based substances was introduced. These industrial biolubricants contain 95% renewable sources and are eco-friendlier with renewable penetrating properties. LCA reports have confirmed that industrial biolubricants contribute to lower carbon footprints than fossil-based lubricants with extremely low emissions of volatile organic compounds and are nonflammable in nature. The unique penetrating nature of industrial biolubricants finds application in corroded machinery parts and machinery that demand light lubrication-. As compared with fossil-based lubricants, the LCA outcomes of industrial biolubricants show beneficial properties such as the reduction in the application of fossil oils, climate-friendlier, environmentally friendlier, long-term use, nontoxic, fragrance-free, and production from used oils and fats, leading to manifold applications.

11.5 Biolubricants: issues, challenges, and conclusion

Nowadays, biolubricants are preferred [12, 13], but there are both advantages and disadvantages. The issues and challenges of biolubricants are high VI, decreased flowability, high flash point, hard storability, low oxidative stability, unpleasant odor, poor compatibility, clogging properties, high cost, high melting points, less developed, and high acidity. Life cycle analysis provides indicators for the development of novel chemical modifications of these properties and helps produce biolubricants with high lubricity, high volatility, lower emissions, longer tool life, minimum release of contaminants, customizable chemical structures, and large availability of plant sources [14].

The demand for biolubricants is predicted to increase steeply in the near future in applications such as hydraulic oils, engine oils, gear oils, cutting fluids, electrical appliances, and turbomachinery. The key challenges involved and the scope for further investigation in the area of biolubricants obtained from overall wastes, biomass residue, plastic wastes, waste cooking oil, agro-residues, and renewable sources are to be considered. Sustainable biolubricant procedures and formulations using suitable additives need to be achieved to increase our present understanding of various properties and stability of biolubricants. The scope for improvement in the making of biolubricants will focus on: (i) reduction in oxygen content in the fuel, (ii) increase in storage rate, (iii) high oxidative stability, (iv) high hydrolytic and thermal stability, (v) improved formulations, (vi) expanded LCA reports on biodegradability and ecotoxicity, (vii) improving diverse applications, (viii) techno-economic assessment, (ix) examination on carbon dioxide emissions, and (x) water depletion and fossil depletion. In the future, suitable blending of bio-based lubricants and mineral-based lubricants may turn out to be a sustainable procedure that is yet to be completely revealed [15]. The transformation of nonedible oil into biolubricants by novel chemical modifications such as transesterification to reproduce hydraulic fluids, engine oil, and grease oil may be developed as a sustainable route to biolubes.

References

[1] Mobarak HM, Mohamad EN, Masjuki HH. The prospects of biolubricants as alternatives in automotive applications. Renewable and Sustainable Energy Reviews. 2014;33:34–43. https://doi.org/10.1016/j.rser.2014.01.062

[2] Dahiya S, Katakojwala R, Ramakrishna S, Mohan SV. Biobased products and life cycle assessment in the context of circular economy and sustainability. Materials Circular Economy. 2020;2:7.

[3] Marcin S, Tomasz T, Krzysztof Ż, Robert O, Marek Z. Effect of lubricant type on the friction behaviours and surface topography in metal forming of Ti-6Al-4V Titanium Alloy Sheets. Materials (Basel). 2021;14(13):3721. https://doi.org/10.3390/ma14133721

[4] Muthusamy B, Subramaniapillai N. A novel biobased heterogeneous catalyst derived from Musa acuminata peduncle for biodiesel production – Process optimization using central composite design. Energy Conversion and Management. 2019;189:118–31. https://DOI:10.1016/j.enconman.2019.03.085

[5] Jumat S, Nadia S, Emad Y. Industrial development and applications of plant oils and their biobased oleochemicals. Arabian Journal of Chemistry. 2012;5(2):135–45. https://doi.org/10.1016/j.arabjc.2010.08.007

[6] Sukirno, Rizqon F, Setijo B, Mohammad N. Biogrease based on palm oil and lithium soap thickener: Evaluation of antiwear property. World Applied Sciences Journal. 2009;6:401–07. http://www.idosi.org/wasj/wasj6(3)/14.pdf

[7] Piyush SL, Bo M. Green approach for the preparation of biodegradable lubricant base stock from epoxidized vegetable oil. Applied Catalysis B: Environmental. 2007;69:207–12. https://doi.org/10.1016/j.apcatb.2006.06.016

[8] Barbera E, Hirayama K, Maglinao RL. Recent developments in synthesizing biolubricants – A review. Biomass Conversion and Biorefinery. 2022. https://doi.org/10.1007/s13399-022-02438-9

[9] Bryony C, Nigel B, Walter W. A sustainability assessment of a biolubricant. Journal of Industrial Ecology. 2003;7(3–4):1–244. https://doi.org/10.1162/108819803323059460.

[10] Mobarak HM, Edzrol NM, Masjuki HH, Ashraful A. The prospects of biolubricants as alternatives in automotive applications. Renewable and Sustainable Energy Reviews. 2014;33:34–43. https://doi.org/10.1016/j.rser.2014.01.062

[11] Reinout H, Gjalt H, Jeroen BG. Life cycle assessment and sustainability analysis of products, materials and technologies. Toward a scientific framework for sustainability life cycle analysis. Polymer Degradation and Stability. 2010;95(3):422–28. https://doi.org/10.1016/j.polymdegradstab.2009.11.010

[12] Rakesh N, Dasappa S. Carbon deposition on the anode of a solid oxide fuel cell fueled by syngas – a thermodynamic analysis. In: Proceedings of the 7th International Conference on Advances in Energy Research, Springer Proceedings in Energy. 2021, 1083–90. ISBN 9789811559549.

[13] Rudnick LR. Synthetics, Mineral Oils, and Bio-Based Lubricants: Chemistry and Technology. 2020. ISBN 978-1-315-15815-0.

[14] Athaley A, Saha B, Ierapetritou M. Biomass-based chemical production using techno-economic and life cycle analysis. AIChE Journal. 2019;65:16660. https://doi.org/10.1002/aic.16660

[15] Rakesh NS, Ravikrishnan V. Current status and future prospects of biolubricants: Properties and applications. Lubricants. 2022;10(4):70. https://doi.org/10.3390/lubricants10040070

Rekha Goswami, Barkha Kama, Abhilasha Mishra

12 Nanotechnology for enhancement of bio-based lubricant production

12.1 Introduction

To uplift the nation's economic development, usage of energy plays a vital role. As per observation by the European Commission, it is predicted that the ingesting of energy in the whole world is to expected to reach 22.3 Gtoe (gigatons of oil equivalent) by 2050 [1] from the current 10 Gtoe. Globally fossils fuels are majorly fulfilling the energy demand.

In modern machinery, lubricants play a vital role, as they have the tendency to lessen the friction and wear between several components present in machines and ease their mobility. Keeping in mind machinery reliability, efficiency, and its life span, various lubricants are used. Lubricants are used in open and closed systems. Comparatively, an open system is much more hazardous than a closed one, because in an open system, the lubricants are expelled to the environment directly, whereas in a closed system the lubricants do not in direct contact with the environment. In general, lubricants are thrust out into the environment due to leakages, human errors, and blown pipes. This results in environmental pollution issues [2, 3]. To avoid the environmental damages, it is necessary to develop lubricants that are biodegradable, renewable, and sustainable. A good lubricant is characterized by high flash point, high viscosity index, better corrosion resistance, high oxidation stability, and low pour point.

From a sustainability point of view, bio-based lubricants have come into attention due to their friendliness toward the environment, as they are derived from renewable raw materials. The sustainability of bio-based lubricants can be evaluated on the basis of two perspectives, the raw materials used and the pollution levels caused by them.

In the context of bio-based products, specifically oleochemicals, their disposal occurs through multiple paths upon reaching the end of their life cycle, resulting in the decomposition of organic molecules into carbon dioxide and water. The carbon cycle of oleochemicals can be considered closed as the quantity of carbon dioxide emitted during their production and consumption is balanced by the amount of carbon dioxide initially absorbed by plants from the environment. Consequently, it exerts no discernible impact on the atmospheric carbon dioxide equilibrium. In contrast, it can be observed that products containing mineral oil contribute to the release of atmospheric carbon dioxide, thus contributing to the phenomenon of global warming. This can be classified as a form of indirect environmental contamination.

https://doi.org/10.1515/9783111145136-012

12.2 Bio-based lubricants sources

Both edible and nonedible oils are considered to be different types of vegetable oil. In spite of the fact that edible oils would be excellent candidates for bio-lubricants, these oils continue to find widespread application in the kitchen, particularly in less developed nations. On the other hand, nonedible oils are now being utilized in the production of soaps, detergents, cosmetics, and other similar items in the chemical industry; nevertheless, the current utilization rate is rather low. This suggests that there is a significant untapped potential for the use of oils derived from non-vegetable sources as lubricants. As a result, they are potential candidates for use in lubricants of the future.

The selection of vegetable oils for the production of bio-based lubricants may vary across countries as a result of variations in meteorological and geographical conditions. In Europe, rapeseed and sunflower oils are frequently employed, while soybean oils are predominantly utilized in the United States. On the other hand, it should be noted that the primary sources of bio-based lubricants in the Asian region predominantly consist of palm and coconut oils. In contemporary times, there exists a growing preference for the utilization of nonconsumable oils in the production of bio-based lubricants, owing to their derivation from agricultural byproducts [4]. Consequently, this practice obviates the necessity of utilizing food crops for the purpose of producing lubricants. Nevertheless, the utilization of nonconsumable oils in the production of bio-based lubricants is advantageous solely under the condition that there exists ample land space to cultivate both edible and nonedible crops. Furthermore, economic considerations may serve as a motivating element for farmers to transition from cultivating food crops to biofuel crops, even if the latter are not suitable for consumption. The above scenario is likely to result in an imbalanced market due to the anticipated rise in food costs, consequent to a decline in food production [5].

12.2.1 Sunflower oil

Sunflowers thrive in soil that is rich in nutrients, adequately moist, and well drained, preferably with the addition of a substantial layer of mulch. Sunflower oil is derived from the seeds of the sunflower plant, and it is frequently employed in culinary applications, including cooking, margarine production, and biodiesel synthesis. According to a study conducted by [6], sunflower oil is cost-effective in comparison to olive oil. There is considerable variation in the fatty acid composition across different sunflower kinds [7]. Certain sunflower varieties, known as "high-oleic" forms, exhibit a higher concentration of monounsaturated fatty acids in their oil when compared to olive oil. High-oleic sunflower oil possesses several characteristics that make it well suited for use as a lubricant, including favorable attributes such as high oxidation stability and lubricity [8, 9]. A study conducted by researchers demonstrated that the utilization of high-oleic sunflower oil (HOSO) as a substitute for mineral oils in textile and tannery applications does not pose any technical challenges or require modifications to existing facilities [10]. The sunflower formulation exhibits equivalent properties to mineral oil, therefore rendering it as suitable substitute lubrication for chainsaws [11].

12.2.2 Rapeseed oil

Rapeseed, a vibrant yellow blooming plant belonging to the mustard family, is characterized by its substantial erucic acid content, which often reaches approximately 45%. The production of rapeseed oil originated in the nineteenth century, primarily serving as a fundamental component for lubricating steam engines. Rapeseed is composed of four primary ingredients, namely, oil, water, protein, and fiber. Rapeseed oil is deemed unsuitable for consumption by both humans and animals due to its bitter taste, primarily attributed to elevated concentrations of glucosinolates [12]. Nevertheless, recent advancements in rapeseed breeding have resulted in the development of novel cultivars, characterized by reduced levels of glucosinolates, rendering them more suitable for consumption.

12.2.3 Canola oil

Canola oil, derived from the seeds of the canola plant, is a widely used vegetable oil in various culinary applications. Canola, a genetically modified cultivar of rapeseed, has received significant commercial recognition, mostly due to its enhanced nutritional attributes, including its decreased levels of erucic acid and glucosinolate content in rapeseed oil. Both canola and rapeseed oils are predominantly made up of unsaturated fats. However, there is a notable distinction in the composition of both oils, as canola oil contains an extremely low level of erucic acid, measuring less than 1% [13].

12.2.4 Soybean oil

The soybean, scientifically known as *Glycine max*, is a leguminous plant species that originates from East Asia. The plant is extensively grown for the purpose of cultivating its beans, which has a multitude of practical applications. Successful production of soybean is observed in regions with hot summers, when the optimal mean temperature falls within the range of 20–30 °C. Soybean cultivation exhibits adaptability to various soil types, while it attains optimal growth on soils characterized by adequate moisture levels, muddy consistency, and favorable organic matter content. The utilization of soybean oil in the production of dielectric liquids for transformers has proven to be effective [14, 15].

12.2.5 Palm oil

Palm oil is a type of consumable oil that is extracted from the mesocarp of the oil palm fruit. The oil palm plant demonstrates optimal growth in tropical climates characterized by high temperatures and humidity, where the annual precipitation falls within the range of 1,500–2,000 mm. Oil palm cultivation exhibits optimal growth conditions when subjected to a maximum of three consecutive months of drought. The optimal temperature range for optimal conditions is observed to be between 26 and 32 °C, while 5–7 h of direct sunlight per day has been found to be advantageous. According to a study [16], the oil palm cultivation on a one-hectare land area yields about ten times more oil in comparison to other types of vegetable oil. Therefore, palm oil exhibits significant promise in meeting the market demand for lubricants derived from vegetable oil.

12.2.6 Coconut oil

The presence of coconut trees in arid locations is infrequent due to their dependence on regular irrigation for growth and survival. The distribution of these trees is primarily concentrated in tropical climates. The fruits produced by coconut trees possess characteristics that make them suitable for dispersal over long distances in sea currents. These fruits are lightweight, buoyant, and exhibit a high degree of water resistance [17]. Coconut oil exhibits a high concentration of saturated fatty acids, comprising approximately 91% of its composition. Consequently, this characteristic renders it less prone to oxidation. The utilization of coconut oil as a lubricant in rickshaws and scooters in Southern India has been prevalent. Research has demonstrated that the utilization of this oil has the potential to enhance vehicle fuel efficiency, engine acceleration, and overall performance. Furthermore, it has been observed that the combustion of coconut oil results in a reduced emission of smoke, as indicated by previous research [18].

12.2.7 Jatropha oil

Jatropha curcas, a member of the Euphorbiaceae family, is a resilient shrub or tree that exhibits drought-tolerant characteristics. Its cultivation is prevalent across several regions, including Central America, South America, Southeast Asia, India, and Africa. Despite being widely used as a feedstock for biodiesel production [19, 20], the lubricating properties of Jatropha oil remain little understood. The high concentration of fatty acids (61–64%) in Jatropha oil renders it a promising candidate for the manufacture of lubricants [21]. The investigation of manufacturing methodologies and optimal parameters required for achieving high yields of lubricant from Jatropha oil is an ongoing research endeavor [22–24].

12.2.8 Castor oil

Ricinus communis, often known as castor, is a perennial shrub, characterized by rapid growth and the ability to produce suckers. It has the potential to attain the stature of a small tree, typically reaching a height of approximately 12 m. However, it should be noted that this plant is not well suited for cold climates due to its lack of cold-hardiness. This particular botanical specimen is commonly referred to as Ricinus communis, or more colloquially, castor beans. The seeds possess an oil content ranging from around 40% to 60%, which is abundant in triglycerides, specifically ricinolein. According to previous research [25], it has been observed that castor oil exhibits superior low-temperature viscosity and high-temperature lubricating characteristics in comparison to the majority of vegetable oils. Hence, castor oil is considered a favorable choice for application as a lubricant in jet, diesel, and race car engines. Castor oil possesses distinctive attributes due to its high ricinoleic acid concentration of 80%, making it the sole provider of C18:1-OH. The oil possesses a distinctive amalgamation of physical characteristics, including a highly elevated viscosity and specific gravity, and the ability to dissolve in alcohol in varying quantities. Nevertheless, the solubility of castor oil in aliphatic petroleum solvents is restricted.

12.2.9 *Calophyllum inophyllum* L.

Calophyllum inophyllum L., also referred to as polanga or honne, is a widely prevalent plant species, primarily distributed along the coastal areas of the Philippines. The plant in question is indigenous to the tropical regions of Asia [26]. The tree in question is characterized by its medium-to-large size, typically reaching heights ranging from 8 to 20 m (25 to 65 feet). It possesses a wide-spreading crown composed of irregularly arranged branches. *Calophyllum inophyllum* L. is a perennial arboreal species that thrives in sub-maritime environments, exhibiting optimal growth on deep soils or exposed sea sands [27]. The seeds of *Calophyllum inophyllum* L. are characterized by substantial oil content, ranging from 65% to 75%. The utilization of *Calophyllum inophyllum* oil extends beyond its medical and hair care applications, as it is also employed in the production of biodiesel that meets the physicochemical requirements outlined by ASTM D 6751 and EN 14214 standards.

In context with global market, the demand for biolubricants is increasing as a prominent substitute for marine, engine, and hydraulic oil, out of which approximately 56% biolubricant is required only for engine oils. Researchers in India, found that biolubricants possess one of the best substitute qualities required for various tribological applications, due to its eco-friendly behavior [28]. One of most important aspects about bio-based lubricants is that they seem to be a capable substitute over conventional ones with regard to technical specifications. Edible and nonedible vegetables oils are the major sources of bio-based lubricants, which show elevated viscosity index, high flash point, and high lubrication quality [29], though it also shows some drawback such as low cold flow and less oxidation stability, which results in polymerization and then degradation. To overcome this problem, the sources used for lubrications are further chemically modified. The main features that enhance the tribological characteristics during the lubrication process are polarity, carbon chain length, and type of fatty acid used.

Since the last few years, ingression of nanomaterials as lubrication additives has become an imperative research field [30–37]. Various nanomaterials are used during lubrication formulation, which enhances its working efficiency in terms of antiwear and antifriction characteristics. Additionally, on adding nanomaterials, it provides a large surface area due to the nanosize. Lubricants also get some new traits such as friction modifiers, extreme pressure tolerance, low melting point, high chemical reactivity due to the addition of nanomaterials and it develops a self-repairing role in addressing the microdefects on friction surfaces [38]. The use of nanomaterials in the lubrication process has shown sustainability due to its environmentally friendly behavior and low usage of hazardous materials. Also, they reduce the emission of carbon due to less energy requirement during the production process [39–41]. The lifecycle of bio-based lubricants and the properties of good lubricants are shown in Fig. 12.1.

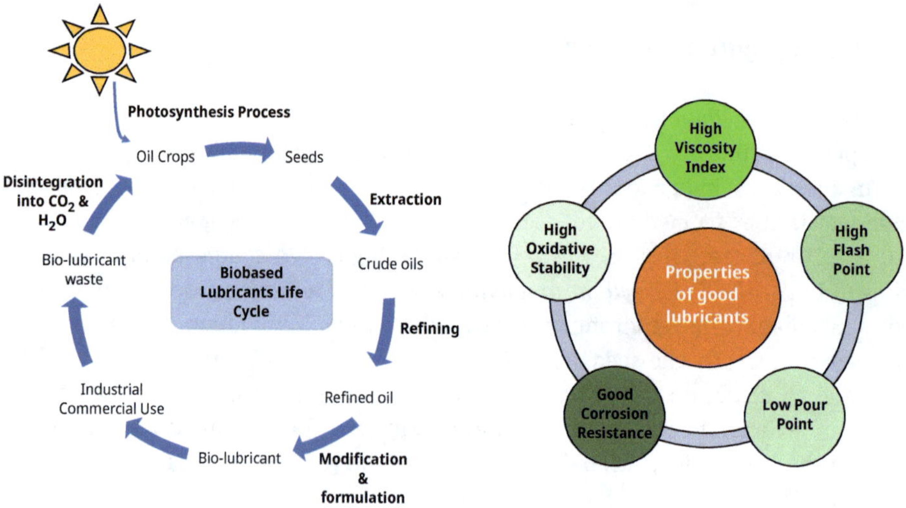

Fig. 12.1: Life cycle of bio-based lubricants and properties of good lubricants.

12.3 Lubrication mechanism of lubricants having nanoparticle additives

Several methods of analyzing surface properties have been used [42–47] in order to explore the lubrication processes of lubricating oils that contain nanoparticles (NPs). This was done in order to learn more about how these oils work. However, it has been demonstrated that the surface analysis methods that are currently in use are not sufficient for determining the specific contribution of NPs in a variety of lubrication mechanisms due to the presence of nanolubricants. The mechanisms of lubrication include a range of different combinations, some of which require the direct action of NPs (such as the formation of ball bearings or a tribofilm), while others involve secondary surface improving processes (like mending or polishing).

12.3.1 The ball bearing effect

In general, NPs have spheric and quasi-spheric aspects as miniature structures like all bearings that roll over the connected area and transform sliding friction into a combined rolling and sliding friction. To be more specific, on adding sphere-shaped NPs of copper oxide over the contact surface to create rolling friction, the observed results showed that the tribological traits of the base lubricants get enhanced [48, 49]. Due to potential effect of NPs during rolling actions in between the contact surface, the introduction of a nanolubricant could result in an improved product quality [50]. This would be due to the fact that the surface contact would be prevented.

12.3.2 The rolling bearing effect

Previous studies discovered that SiO_2 NPs behaved like rolling bearings in the presence of friction and had friction-reducing and wear-resistant capabilities [51]. According to the findings of Rapoport et al. [52], fullerene-like WS2 NPs evidently improved the tribological capabilities; this improvement may be due to the gradual release of WS2 NPs from the matrix and their rolling effects on the rubbing surfaces. According to a study by researchers [53], the rolling action of nanodiamond may be responsible for the friction-reducing capabilities that nanodiamond possesses. In addition, polymer NPs have the ability to produce a tribofilm on the contact surfaces, which allows them to perform the same role as ball bearings when the load on the system is minimal. On the other hand, the rolling bearing effect is susceptible to failure [54] when the applied load is of a sufficient magnitude. The tribological properties of hairy silica NPs were examined by Sui et al. [55], and it was discovered that the exceptional friction reduction and wear-resistance performances could be partially related to the rolling bearing effect. According to the findings of the research, a rise in size was similarly associated with an increase in the rolling bearing effect. The researchers Bao et al. [56] discovered that the thickness of the film is a significant component that influences the lubrication mechanisms. NPs, on the other hand, have a tendency to produce a transfer film when the film is thin. Refer Fig. 12.2(A) for rolling bearing effects' property enhancement utilizing NPs.

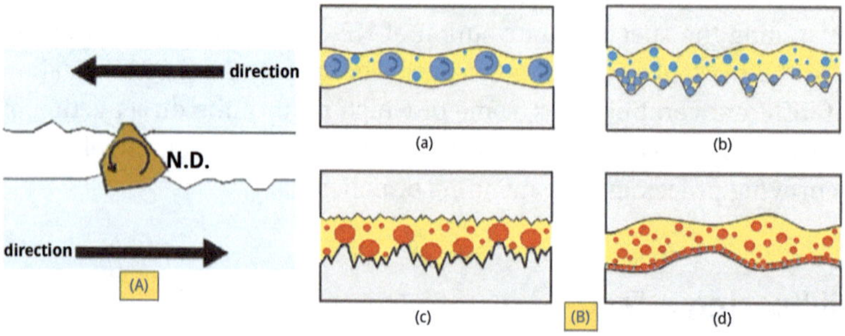

Fig. 12.2: (A) Rolling bearing effect of nanodiamond (reused with permission from [63], @ 2013 Elsevier Ltd.) and (B) Different mechanisms for improving tribological performance using nanoparticles (a) rolling mechanism; (b) mending mechanism; (c) polishing mechanism; (d) protective film (reused with permission from [64], @ 2020 Elsevier Ltd.).

12.3.3 Protective film formation

A tribofilm is another name for the protective film that is placed on the surfaces that are being examined [57]. Tribofilms and other materials are located close to the surface-NP-based ball-bearing system, shown here in Fig. 12.3 to find out how the tribological behavior of friction surfaces should be determined. The creation of a film is kicked off by a reaction involving the treated material and additives that takes place either under the influence of the environment or through tribo-sintering [58].

Experiments have been carried out a number of times to investigate and describe the mechanism of tribofilm production, which is necessary for good lubrication. The mechanical strength and thickness of the tribofilm generated on worn surfaces are both connected with tribological features [59].

It has been established that, in comparison to base oil, the molecules of a nanolubricant become more ordered and condensed as a result of the adsorption layer that is formed around NPs. In addition, the malleable Cu NPs are distorted by the structural elements of the nanolubricant film, which results in a nanolubricant film having a strong support system.

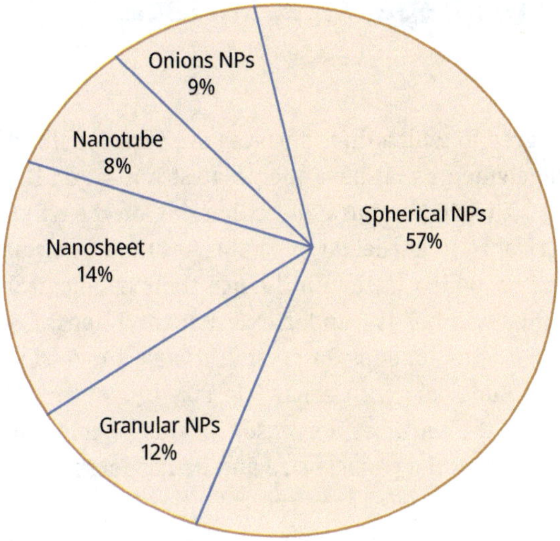

Fig. 12.3: Morphology statistics of nanoparticles.

12.3.4 Mending effect or self-healing effect

This effect, is also known as the self-repairing effect, is distinguished by the deposition of NPs on friction surfaces and the compensation of mass loss. During this phenomenon, NPs deposit on the worn surface and fill the scars and grooves of the friction surface. This deposit helps to prevent abrasion and maintain the surface's smoothness. When compared to conventional lubricants, nanolubricants show a considerable improvement in surface roughness [60]. This was proved by an investigation into the effects that nanolubricants that comprised CuO and Al_2O_3 NPs had on the surface quality of the forged product during the forging process [61]. These results can be attributed to the high load wear index and low cylinder liner wear.

12.3.5 Polishing effect

When it comes to tribological contacts, NPs have the ability to fill the spaces left behind by rough asperities, which can then serve as reservoirs for solid lubricants (NPs). The operation of smoothing out rocky valleys is referred to as the procedure of filling up rough valleys. This method of "artificial smoothing" or polishing leads to improved tribological qualities, principally as a result of reduced surface roughness [62].

12.4 Effect of nanoparticle shape, size, surface, and concentration

NPs' sizes and shapes, as well as the degree to which their surfaces are functionalized and their concentration, are among the elements that have the greatest impact on the tribological properties of nanolubricants. The NP size has a direct bearing on the tribological properties exhibited by nanolubricants. In particular, it establishes their internal mechanical and physicochemical qualities, which in turn influence their tribological features. Additionally, it decides how they will behave under external conditions. Because of an increase in the number of dislocation pileups for crystals that occurs during the Hall–Petch regime, the hardness of materials with a particle size of 100 nm or greater or smaller increases as the size of the particles decreases. In this regime, the relationship between the inverse square root of the particle size and its hardness is linear. In the inverse Hall–Petch regime, nanomaterials reach critical grain sizes, which are typically below 10 nm, and grow softer as their size declines. When the hardness of the NP material is higher than the hardness of the material that it is rubbing against, indentation and scratching might occur.

12.4.1 Shape

Nanolubricant design also depends on the form of NPs utilized as additives, which influences their loading pressure. Five NP forms are granular, onion, sheet, spherical, and tube. After nucleation, crystalline particle architectures develop to minimize surface energy, resulting in a spherical form. Onions have a spherical exterior and are lamellar within. The onion morphology is more spherical, if stable. It will exfoliate and become sheet-like, otherwise. The onion structure's spherical shape and lack of hanging connections are benefits. Onions, leaves, and spherical morphologies are tribologically superior. Spherically shaped NPs have high load capacity and EP properties because of their ball bearing effect, which reduces friction by rolling instead of sliding [65, 66]. The spherical form of NPs makes point contact with the counter. Line contact is seen in nanosheets and planar contact in nanoplatelets.

The tribological properties of single-crystalline α- and β-MnO_2 nanorods as nanoadditives in green lubricants are of interest [67]. The friction torque was lowest for α-MnO $_2$-added palm oil, followed by β-MnO_2-added palm oil and pure palm oil (Fig. 12.4(a)). Despite a 15% coefficient of friction (COF) reduction, α-MnO_2 nanorods are significantly more effective, reducing COF by up to 30% (Fig. 12.4(b)). Antiwear capability increases due to the rolling motion and due to the development of the quasi-1D MnO_2 polymorphs' protective layer.

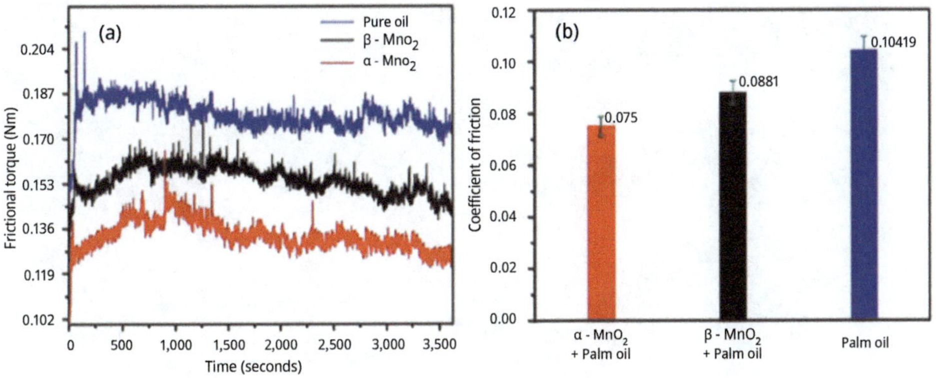

Fig. 12.4: (a) Frictional torque with respect to time observed using four-ball test technique and (b) COF of pure palm oil, β-MnO_2-added palm oil, and α-MnO_2-added palm oil (reused with permission from [68], @ 2019 Springer Nature).

12.4.2 Size

Nanolubricants' tribology depends on the NP size. It affects their tribological qualities by determining their internal mechanical and physicochemical properties. For materials 100 nm or larger, hardness improves as particle size decreases due to crystal dislocation pileups (Hall–Petch regime). Hardness rises linearly with inverse square root of particle size in this domain. Nanomaterials soften when grain size falls, generally below 10 nm (inverse Hall–Petch regime). Indentation and scratching arise from NP hardness exceeding tribo-pair material hardness. The increased hardness (8–9 Mohs) of nano-Al_2O_3, relative to metal substrates, causes NP abrasion and re-agglomeration [69]. Thus, NP size and hardness must be considered while designing a nanolubricant. The ratio of the lubricating surface's root-mean-square roughness to the NP radius is crucial when choosing an NP size. To safeguard friction surfaces, NP-based lubrication systems must stay in the contact zone during loading and shearing. NPs will not deposit on the contact zone if their size exceeds the typical roughness length scale of the shearing surfaces, resulting in inadequate lubrication. The troughs between shearing surface asperities can be filled by NPs when the characteristic roughness length scale is much bigger than the NP radius. Finally, colloidal stability determines the nanolubricant position's homogeneity, which affects tribology. NP size determines dispersion stability, which is crucial for nanolubricant composition. Refer Fig. 12.5(a) and (b) for the impact of NPs on the lubrication properties.

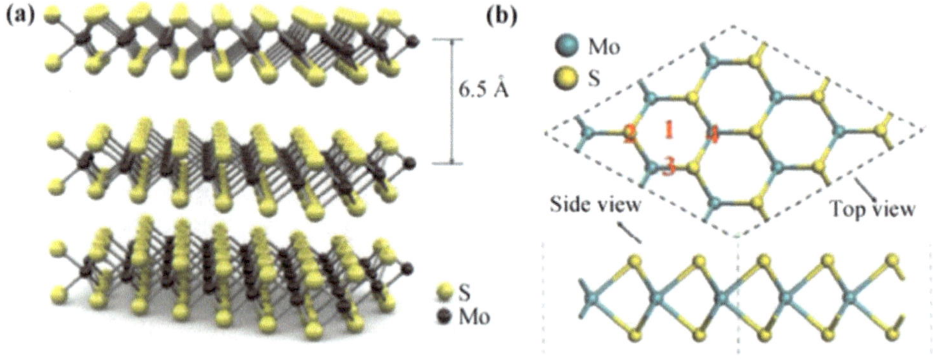

Fig. 12.5: (a) Three-dimensional representation of the structure of MoS_2. Single layers, 6.5 Å thick, can be extracted using scotch tape-based micromechanical cleavage. (b) Optimized structures of MoS_2 monolayer with four adsorption sites: (1) hollow site, (2) top site of the S atom, (3) Mo-S bridge site, and (4) top site of the Mo atom (reused with permission from [70], @ 2016 Elsevier Ltd.).

12.4.3 NP concentration

Another important factor affecting nanolubricant lubrication is concentration. NPs reduce friction and wear even at concentrations below 1 wt% [71] and above 2 wt% [72], indicating that there is no optimal concentration. Concentration does not predict nanoadditive friction and wear effects. Note that COF is lowest at a certain concentration. Because each operating condition requires a different lubricant composition, it is highly system-dependent [73]. The same MoS_2 NPs work at 0.58 wt% for mineral oil and 0.53 wt% for coconut oil [74]. For mineral, synthetic, and vegetable oils, 0.5 wt% CuO and ZnO NPs was optimal [75]. Research on the tribo-performance of cerium oxide (≈90 nm) NPs in paraffin oil found that 0.25% w/v is optimal for antiwear and antifriction tests [71]. The maximum wear scar diameter (WSD) reduction was 26.1% and the average COF reduction was 29.6% at this concentration. Nanolubricants based on sunflower oil and two nanoadditives, CuO and CeO_2, with concentrations from 0.10% to 0.50% w/v, have the least wear scar and COF at 0.10%. High NP concentrations degrade base oil performance. At 0.42% Hf, diamond-like thin films doped with hafnium had low COF and excellent wear resistance [76].

NPs may reduce boundary lubrication friction by reducing the real area of contact [77, 78]. Thus, the particles in contact will keep the surfaces apart, decreasing the real contact area and COF. The monotonic decrease in COF compared to particle concentration in the tests is due to more particles coming into contact with more concentrated nanolubricants. The results show that each type of NP should be analyzed based on size, morphology, surface functionalization, concentration, application conditions (temperature, load, and slip speed), and base lubricant.

12.4.4 Surface functionalization

Functionalizing the NP surface regulates colloidal stability and increases lubricity in most NP layers. Due to material transfer and cold-welding prevention, bare NPs have poor lubrication compared to functionalized ones. Functionalized NPs also prevent material transfer and cold-welding between shearing surfaces. The hybrid structure of functionalized NPs with a rigid inner core and soft outer shell is crucial. This synergistic combination gives NPs a rigid internal shape and a slippery fluid-like surface. Such NPs increase load capacity without reducing lubricity [79]. Surfactant-covered finely dispersed Cu NPs were added to engine oils [80]. Creating protective films on metal surfaces involves tribological processes like absorption of polar molecules and mechanochemical processes like redox reactions and third body formation to form an FM film. Oil-soluble Cu NPs, modified with tetradecyl hydroxamic acid, are environmentally friendly oil additives that improve antiwear and reduce friction. Under higher loads, Cu NPs deposit and fill micropits and grooves on steel friction surfaces, self-repairing worn surfaces and reducing steel pair wear [81]. Triangular copper nanoplates are capped with cetyltrimethylammonium bromide [82].

Nanoplates in lubricants form a film deposit at the friction pair interface, reducing COF by 12% and wear loss by 82.2%. Dioctylamine dithiocarbamate [83] or alkanethiols [84]-capped Cu NPs were added to liquid paraffin. Due to the deposition of Cu NPs with a low melting point on the worn steel surface, they form a self-repairing protective layer that reduces friction and wear.

12.5 Types of nanoparticles used as additives in lubricants

Nanomaterials are known to be one of the fascinating technologies used in the domain of material science, chemistry, and physics. Nanomaterials as lubricant additives can form a protective tribofilm and reduce friction contact areas due to their small size, preventing friction pair wear. Nanomaterials' high surface activity improves the protecting tribofilm film-forming stability through physical and chemical adsorption. Nanometallic, nanocarbon, and nanocomposite nanolubricant additives exist.

With the progress of nanotechnology, its associated nanomaterials have gained immense attraction in the domain of physics, material sciences, and chemistry. In a lubrication system, nanomaterials act as an additive due to their tiny sizes and high availability of surface, which easily enter the friction contact areas and develop a protective tribofilm, preventing the surface from any damage. Generally, there are three types of nanoadditives used in lubricants. They are listed as metal-based, carbon-based, and composite-based nanoadditives.

12.5.1 Metal-based lubricant nanoadditives

This class of additive includes metal oxides, metals, metal hydroxides, metal sulfides, and metal salts. In the 1980s, researchers initially used copper NPs as lubricant additives in base oils, which resulted in improved tribological characteristics.

Recent studies demonstrate the enhanced tribological characteristics result from the incorporation of metallic NPs, including but not limited to Fe, Pd, and Ni [84–88]. Comparative investigations have revealed that metal oxides, including ZnO [89, 90], TiO_2 [91–93], and ZrO_2 [94], have favorable lubricating properties. Furthermore, the characteristics of nanometal oxides are intricately linked to factors such as particle size, hardness, and concentration.

According to the findings of Wu et al. [95], the inclusion of CuO nanosheets as additives in self-lubricating carbon fibers, reinforced with polytetrafluoroethylene (PTFE), resulted in a noticeable improvement in antiwear performance. Additionally, the study also discovered that CuO nanosheets demonstrated enhanced antiwear characteristics, compared to commercially available CuO nanogranules. Furthermore, there has been significant interest in the subject of lubrication about the utilization of metal sulfides as additives. For instance, the chemical interaction between MoS_2 and tungsten disulfide (WS_2) and friction pairings might result in the formation of a durable adsorption film on the surfaces experiencing rubbing. This phenomenon can substantially enhance the tribological characteristics of lubricants, particularly in situations involving intense friction circumstances [96].

In a recent study conducted by Yi and Zhang [97], it was shown that MoS_2 with a lamellar morphology exhibited significantly superior tribological capabilities compared to its petal structure counterpart. This finding suggests that MoS_2 with a lamellar structure possesses heightened chemical activity and is more prone to displaying interlayer sliding effects. Furthermore, the study conducted by Hu et al. [98] provided confirmation that spherical MoS_2 NPs exhibited a more pronounced antioxidant impact compared to MoS_2 NPs characterized by a layered structure. This enhanced antioxidant effect subsequently contributed to the improvement of lubrication stability.

Furthermore, it has been demonstrated that metal hydroxides have exceptional lubricating capabilities. Among them, layered double hydroxides (LDHs) stand out as the most prominent examples. LDHs powders demonstrate exceptional friction and wear reduction capabilities, particularly in situations of boundary lubrication, due to their laminated structure, tiny particle size, and strong chemical activity [99, 100]. In addition to LDHs, $La(OH)_3$ NPs, which belong to a distinct category of metal hydroxides, have also demonstrated efficacy as lubricant additives. In their study, Zhao et al. [101] employed the sol-gel method to synthesize $La(OH)^3$ NPs, characterized by granular morphologies. Researchers subsequently conducted experiments to validate the improved friction-reduction and wear-resistance capabilities of base oil upon the incorporation of $La(OH)_3$ NPs. Metal salts have the capacity to serve as additives in lubricants. The utilization of $CaCO_3$ NPs as additives for lubricants was subject to investigation throughout the early 2000s. Research findings indicate that the utilization of $CaCO_3$ NPs can effectively enhance lubrication by promoting the creation of a tribofilm [102].

Furthermore, the research conducted by He et al. [103] substantiated the exceptional antifriction capabilities exhibited by α-Zirconium phosphate (ZrP) nanoplatelets when employed as additives in lubricants. The findings indicated a reduction in the friction coefficient of 65% and 91% when compared to mineral oil and water, respectively. The exceptional reduction in friction seen in the study was ascribed to the alteration of viscosity and intermolecular interactions. Previous research has also investigated the tribological properties of calcium borate [104], zinc phosphate [105], and LaF_3 [106].

12.5.2 Carbon-based nanoadditives

Chemically, carbon-based nanoadditives consist of pure carbon and polymers. Carbon-based nanoadditives are sustainable in nature and, comparatively, show exceptional tribological traits due to their enhanced stability and mechanical behavior. Moreover, on the basis of the different dimension areas, carbon-based additives include classes of carbon quantum dots, fullerenes, graphenes, graphites, carbon nanotubes, and nanodiamonds [107].

Researchers compared the tribological properties of nanodiamonds and fullerenes as lubricant additives. Observed results showed that the nanodiamond lubricant additives are inclined toward surface abrasion when subjected to sliding motion. On the other hand, fullerenes showed good wear resistance due to their better shear deformation traits. Studies showed that after surface functionalization over carbon nanotubes, they show a good dispersity index and micro bearing rolling effect during the lubrication mechanism [108, 109].

Additives such as PTFE NPs were greased over synthesized lithium in a previous study and the study revealed that they enhanced properties such as friction reduction and wear resistance. Observed results also revealed that size and shape are the two factors that affect the lubrication mechanism [110]. To enhance the antifriction and wear resistance properties in lubricants, researchers studied the tribological properties of polymeric NPs such as poly (sodium 4-styrenesulphonate) (PSS) and polyvinylpyrrolidone (PVP) [111, 112].

In their study, Eswaraiah et al. [113] employed concentrated solar radiation to exfoliate ultrathin graphene from graphite oxide. They subsequently examined the tribological characteristics of graphene, when utilized as additives. The findings demonstrated significant enhancements in the qualities of friction reduction, antiwear, and extreme pressure, with improvements of 80%, 33%, and 40% respectively. In their study, Mungse et al. [114] successfully synthesized graphene by employing a chemical reducing agent to reduce graphene oxide. Their findings demonstrated that the incorporation of graphene led to a substantial improvement in the tribological properties.

The tribological performance of thermally reduced graphite oxide was investigated by Schlüter et al. [115], who verified its exceptional characteristics in reducing friction and enhancing wear resistance. The reduction of graphite oxide offers operational and cost benefits; however, it should be noted that the reduction process can result in the formation of structural flaws in graphene [116]. During the process of thermal reduction, liberation of gases leads to the formation of voids and deformations on the graphene matrix, thus influencing its lubrication capabilities. Zhao et al. [117] discovered that the introduction of certain chemical agents, such as H_2SO_4 and KOH, can effectively impede the occurrence of structural flaws in graphene during thermal reduction. This finding holds significant implications for the production of graphene with exceptional antifriction and antiwear characteristics.

12.5.3 Nanocomposite-based lubricant additives

Comparatively, nanocomposite-based additives show outstanding lubrication mechanism due to the formation of a composite tribolayer over the friction interfaces, showing synergistic effects. Previous studies showed that nanocomposites such as $Cu@SiO_2$ and $Al_2O_3@TiO_2$ used as lubricant additives results in enhanced tribological characteristics [118, 119]. To enhance the friction reduction, antiwear properties, and load-carrying frequency of the lubricants, various nanocomposites such as PVA@CNTs and multiwalled carbon nanotubes were synthesized in previous studies [120, 121]. Nanocomposite-based additives ($Cu@MoS_2$) were synthesized and investigated for tribological properties. When compared with pure sunflower oils, the observed results showed that the friction coefficient and wear scar were decreased by 38% and 29% in the case of $Cu@MoS_2$ and pure sunflower oils, respectively [122]. Song et al. [123] successfully synthesized MoS_2@graphene (MoS_2@G) composites. Their work subsequently explored the tribological performance of these composites. Research indicates that the antifriction and antiwear characteristics of MoS_2@G nanocomposites have the potential to be greatly enhanced. The production of nanocomposites is accomplished by the use of the hydrothermal process, majority of the time. It is simple to exert control over the shape of NPs and their distribution density on nanocomposites. Control can also be exercised over the NPs' location.

Previous examined studies showed that on combining two NPs to form a nanocomposite, that is, α-Fe_2O_3@GO, the lubrication mechanism is enhanced in terms of tribological properties, and gives better results in the composite form rather than individual use of Fe_2O_3 NPs and graphene. Graphene composite with other NPs such as silver, copper, and FeS_2 also shows enhanced antifriction as well as antiwear mechanism [124–127].

Specifically, Zhao et al. [128] used in situ green methods to synthesize Mn_3O_4@G nanocomposites. The test findings revealed that the friction coefficient and wear depth were reduced by 75% and 97%, respectively, while using the additive, in comparison to the base oil, at a very low concentration (0.075 wt%). This was in comparison to the fact that the additive was used. Nanolubricant additives are quickly becoming the industry standard due to their better antifriction and wear-reduction properties, which are causing traditional lubricant additives to be phased out. Due to these distinguishing characteristics, nanolubricant additives stand out as the preferred option.

12.6 Nanomaterial design categories for lubricating oil additives

Nanomaterials belong to a class having dimensions in the range of 1–500 nm. Dispersed in oil, functionalized nanomaterials can pass undisturbed through filters and easily enter the contact interface to reduce friction and prevent friction pairs from being worn due to tribofilm formation, rolling mechanism, or surface repairing [129, 130]. Because it affects loading pressure, nanomaterials' nanostructure – sphere, tube, or layer – is also important. In particular, spherical NPs can withstand large loads and pressures. Categories for nanomaterials design are summarized and illustrated in Fig. 12.6.

It is commonly known that tiny substances have qualities that bulkier ones do not. Most NPs have excellent mechanical strength, load capacity, and specific surface area [131, 132], which improve lubricating oil performance. Nanomaterials, with significant adsorption, can combine various chemicals to produce nanocomposites or disperse them in lubricants due to their huge specific surface areas. A portion of layer nanomaterials with weak interlayer contacts may have low interfacial shear force, reducing friction [133]. However, most nanomaterials have high thermal conductivity, which releases frictional heat and stabilizes friction systems. Nanomaterials are thermally stable at higher temperatures than organic oil additives, which is important for lubricating oil's durability [134].

Nanomaterials can be used as host platforms for functional chemical groups and species to graft selected groups and create composite nanomaterials with improved or new performance [135]. Thus, controlling the size, functionalization, and composition of integrated nanomaterials should allow their important characteristics to be modified systematically [136]. Most non-functionalized NPs aggregate in nonpolar oil [137], which mechanical stirring cannot modify.

Nanocomposites can demonstrate a synergistic effect in thin film boundaries, which might increase nanomaterials' tribological characteristics and be used in many applications [138]. The synergistic lubricating effect makes two or more NPs perform better tribologically than solo nanomaterials. Nanocomposites provide better friction

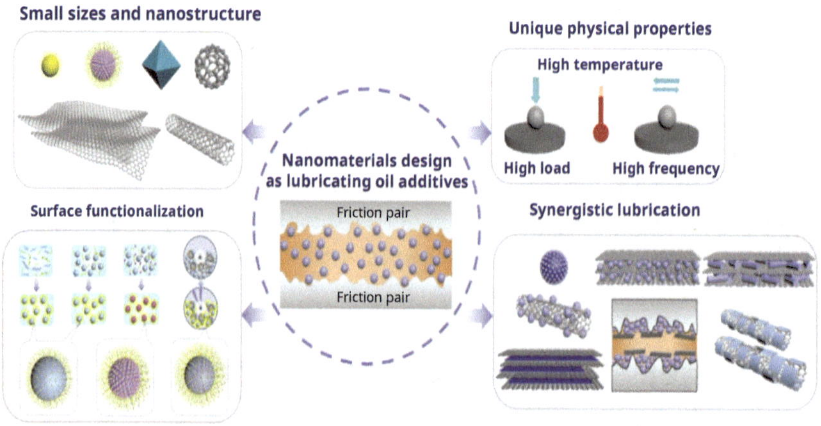

Fig. 12.6: Nanomaterial design categorization as lubricant additives (reused with permission from [141], @ 2023 Springer Nature).

reduction, antiwear, antioxidation ability, oil dispersibility, and load-carrying ability due to synergistic processes [139]. Nanomaterial additions can combine the benefits of organic compounds and solid lubricants [140]. The categories of surface functionalization are multifarious, and primary methods are summarized by four processes, as shown in Fig. 12.7.

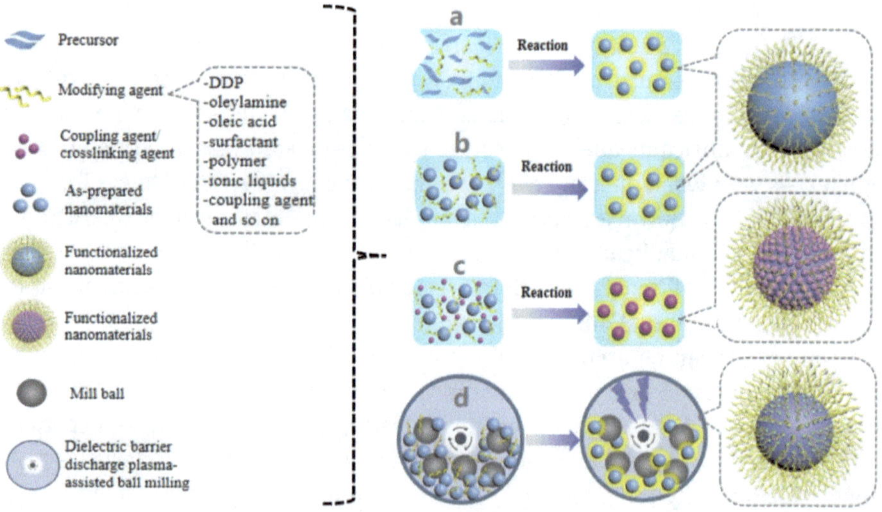

Fig. 12.7: Functionalized nanomaterials' preparation methods: (a) precursors and modifying agent reaction, (b) reaction with modifying agent, (c) use of coupling agent and cross-linking agent, and (d) modifying agent and plasma-assisted ball milling-based reactions. DDP, dialkyldithiophosphate (reused with permission from [141], @ 2023 Springer Nature).

12.7 Conclusion and future prospects

This chapter presents an extensive review of the use of NPs as additives in lubricants, which can be found elsewhere on this website. An exhaustive investigation into the performance characteristics, method of preparation, and working mechanism of nanolubricants is carried out here. Recent studies have shown that the utilization of NPs in lubricants has the potential to bring about a substantial improvement in the products' overall performance. Lubricants have significant improvements in their properties, including their load-carrying capacity, COF, WSD, viscosity, and pressure properties. Consequently, it is imperative that efforts be directed toward the improvement of NP production techniques in order to enhance the economic feasibility of producing NPs. In order to improve the performance of a biolubricant that is primarily based on vegetable oil, it is essential to optimize the NP size and concentration. However, despite the aforementioned factors, lubricants derived from vegetable oils exhibit promising potential in the future due to their ability to address numerous limitations and obstacles associated with traditional mineral or synthetic lubricants.

References

[1] Commission E. World Energy Technology Outlook – 2050 (WETO-H2). Brussels: European Commission, 2006.

[2] Rudnick LR, ed. Synthetics, Mineral Oils, and Bio-based Lubricants: Chemistry and Technology. Laramie, USA: CRC press, 2020 Jan 29.

[3] Willing A. Lubricants based on renewable resources–an environmentally compatible alternative to mineral oil products. Chemosphere. 2001 Apr 1;43(1):89–98.

[4] Atabani AE, Silitonga AS, Ong HC, Mahlia TM, Masjuki HH, Badruddin IA, Fayaz H. Non-edible vegetable oils: A critical evaluation of oil extraction, fatty acid compositions, biodiesel production, characteristics, engine performance and emissions production. Renewable and Sustainable Energy Reviews. 2013 Feb 1;18:211–45.

[5] Kenney VP, Erichsen RL. Conflict between fuel and food: The ethical dimension. Energy in Agriculture. 1983 Jan 1;2:285–306.

[6] Dominguez Brando J, Sarquis A. Challenges for the sunflower oil market for 2020. In: Proc. 18th Int. Sunflower Conf. Mar del Plata-Balcarce, ARG 2012. 35–42.

[7] Putt ED, Carson RB. Variation in composition of sunflower oil from composite samples and single seeds of varieties and inbred lines. Journal of the American Oil Chemists' Society. 1969 Mar;46(3):126–29.

[8] Minami I, Hong H-S, Mathur NC. Lubrication performance of model organic compounds in high oleic sunflower oil. Journal of Synthetic Lubrication. 1999;16:1–12.

[9] Quinchia LA, Delgado MA, Valencia C, Franco JM, Gallegos C. Viscosity modification of high-oleic sunflower oil with polymeric additives for the design of new biolubricant formulations. Environmental Science and Technology. 2009;43:2060–65.

[10] Lazzeri L, Mazzoncini M, Rossi A, Balducci E, Bartolini G, Giovannelli L, et al. Biolubricants for the textile and tannery industries as an alternative to conventional mineral oils: An application experience in the Tuscany province. Industrial Crops and Products. 2006;24:280–91.

[11] Rac A, Vencl A. Performance investigation of chain saw lubricants based on new sunflower oil (NSO). TribolSchmier. 2009;56:51–54.
[12] Cartea M, Velasco P. Glucosinolates in Brassica foods: Bioavailability in food and significance for human health. Phytochemistry Reviews. 2008;7:213–29.
[13] O'brien RD. Fats and Oils: Formulating and Processing for Applications. Boca Raton: CRC press, 2008.
[14] Cannon GS, Honary LAT. Soybean based transformer oil and transmission line fluid. Google Patents, 1999.
[15] Robia Y, Amir N. Use of natural vegetable oils as alternative dielectric transformer coolants. Journal. 2006;67.
[16] Ming KK, Chandramohan D. Malaysian palm oil Industry at crossroads and its future direction. Oil Palm Industry Economic Journal. 2002;2(2):10–15.
[17] Foale M. The coconut odyssey: The bounteous possibilities of the tree of life. Australian Centre for International Agricultural Research, Canberra ACT 2601. 2003.
[18] Jayadas NH, Prabhakaran Nair K, Ajithkumar G. Tribological evaluation of coconut oil as an environment-friendly lubricant. Tribology International. 2007;40:350–54.
[19] Juan JC, Kartika DA, Wu TY, Hin T-YY. Biodiesel production from jatropha oil by catalytic and non-catalytic approaches: An overview. Bioresource Technology. 2011;102:452–60.
[20] Kumar MS, Ramesh A, Nagalingam B. Complete vegetable oil fueled dual fuel compression ignition engine. SAE International. 2001;28:67.
[21] Akbar E, Yaakob Z, Kamarudin SK, Ismail M, Salimon J. Characteristic and composition of Jatropha curcas oil seed from Malaysia and its potential as biodiesel feedstock feedstock. European Journal of Scientific Research. 2009;29:396–403.
[22] Arbain NH, Salimon J. The effects of various acid catalyst on the esterification of jatropha curcas oil based trimethylolpropane ester as biolubricant base stock. European Journal of Chemistry. 2011;8:S33–40.
[23] GunamResul MFM, Mohd Ghazi TI, Idris A. Kinetic study of jatropha biolubricant from transesterification of jatropha curcas oil with trimethylolpropane: Effects of temperature. Industrial Crops and Products. 2012;38:87–92.
[24] Resul G, Mukhtar MF, Ghazi M, Idaty T, Muhammad Syam A, Idris A. Synthesis of biodegradable lubricant from jatropha oil with high content of free fatty acids. 2008 AIChE Annual Meeting, November 16–21, 2008.
[25] Ogunniyi DS. Castor oil: A vital industrial raw material. Bioresource Technology. 2006;97:1086–91.
[26] Venkanna B, Reddy CV. Biodiesel production and optimization from Calophylluminophyllumlinn oil (honne oil)–a three stage method. Bioresource Technology. 2009;100:5122–25.
[27] Atabani AE, Silitonga AS, Ong HC, Mahlia TMI, Masjuki HH, Badruddin IA, et al. non-edible vegetable oils: A critical evaluation of oil extraction, fatty acid compositions, biodiesel production, characteristics, engine performance and emissions production. Renewable and Sustainable Energy Reviews. 2013;18:211–45.
[28] Madankar CS, Pradhan S, Naik SN. Parametric study of reactive extraction of castor seed (Ricinus communis L.) for methyl ester production and its potential use as bio lubricant. Industrial Crops and Products. 2013;43:283–90.
[29] Mobarak HM, Niza Mohamad E, Masjuki HH, Kalam MA, Al Mahmud KAH, Habibullah M, et al. The prospects of biolubricants as alternatives in automotive applications. Renewable and Sustainable Energy Reviews. 2014;33:34–43.
[30] Zhang Y, Chromik R. Self-lubricating composites In: Menezes PL, Rohatgi PK, Omrani E, eds. Cham: Springer, 2018. Vol. 2022. ISBN: 978-3-662-64242-9 (Print) 978-3-662-64243-6 (Online). doi: 10.1007/978-3-662-64243-6.

[31] Tribocatalysis T, Tribocorrosion Kajdas C, Hiratsuka K, eds. Tribocatalysis, Tribochemistry, and Tribocorrosion. New York:CRC Press, Pan Stanford, 2018.
[32] Gnanasekaran D, Chavidi VP. Vegetable Oil-based Biolubricants and Transformer Fluids. Singapore: Springer, 2018.
[33] Darminesh SP, Sidik NAC, Najafi G, Mamat R, Ken TL, Asako Y. Recent development on biodegradable nanolubricant: A review. International Journal of Heat and Mass Transfer. 2017;86:159–65. doi: 10.1016/j.icheatmasstransfer.2017.05.022.
[34] Tang Z, Li S. A review of recent developments of friction modifiers for liquid lubricants (2007-present). Current Opinion in Solid State and Materials Science. 2014;18(3):119–39. doi: 10.1016/j.cossms.2014.02.002.
[35] Xiao H, Liu S. 2D nanomaterials as lubricant additive: A review. Materials and Design. 2017;135:319–32. doi: 10.1016/j.matdes.2017.09.029.
[36] Liu L, Zhou M, Li X, Jin L, Su G, Mo Y, et al. Research progress in application of 2D materials in liquid-phase lubrication system. Materials (Basel). 2018;11(8):1314. doi: 10.3390/ma11081314. PMID 30061482
[37] Shahnazar S, Bagheri S, Abd Hamid SBA. Enhancing lubricant properties by nanoparticle additives. International Journal of Hydrogen Energy. 2016;41(4):3153–70. doi: 10.1016/j.ijhydene.2015.12.040.
[38] Kong L, Sun J, Bao Y. Preparation, characterization and tribological mechanism of nanofluids. RSC Advances. 2017;7(21):12599–609. doi: 10.1039/C6RA28243A.
[39] Peña-Parás L, Maldonado-Cortés D, Taha-Tijerina J. Handbook of Ecomaterials. In: Martínez L, Kharissova O, Kharisov B, eds. Cham: Springer, 2019, XLVIII, 3773. https://doi.org/10.1007/978-3-319-68255-6.
[40] Zhang S-W. Green tribology: Fundamentals and future development. Friction. 2013;1(2):186–94. doi: 10.1007/s40544-013-0012-4.
[41] Zhang S-W. Recent developments of green tribology. Surface Topography: Metrology and Properties. 2016;4(2):023004. doi: 10.1088/2051-672X/4/2/023004.
[42] Padgurskas J, Rukuiza R, Prosycevas I, Kreuvaitis R. Tribological properties of lubricant additives of Fe, Cu and Co nanoparticles. Tribology International. 2013;60:224–32.
[43] Thakre AA, Thakur A. Study of behaviour of aluminium oxide nanoparticles suspended in SAE20W40 oil under extreme pressure lubrication. Industrial Lubrication and Tribology. 2015;67:328.
[44] Alves SM, Barros BS, Trajano MF, Ribeiro KSB, Moura E. Tribological behavior of vegetable oil-based lubricants with nanoparticles of oxides in boundary lubrication conditions. Tribology International. 2013;65:28–36.
[45] Wu L, Zhang Y, Yang G, Zhang S, Yu L, Zhang P. Tribological properties of oleic acid-modified zinc oxide nanoparticles as the lubricant additive in poly-alpha olefin and diisooctylsebacate base oils. RSC Advances. 2016;6:69836–44.
[46] Ettefaghi E, Ahmadi H, Rashidi A, Seyed SM, Mahshad A. Experimental evaluation of engine oil properties containing copper oxide nanoparticles as a nanoadditive. International Journal of Industrial Chemistry. 2013;4:28.
[47] Gulzar M, Masjuki HH, Kalam MA, Varman M, Zulkifli NWM. Antiwear behavior of CuO nanoparticles as additive in bio-based lubricant. Key Engineering Materials. 2017;748:166–70.
[48] Alves SM, Barros BS, Trajano MF, Ribeiro KSB, Moura E. Tribological behavior of vegetable oil-based lubricants with nanoparticles of oxides in boundary lubrication conditions. Tribology International. 2013;65:28–36.
[49] Peña-Parás L, Taha-Tijerina J, García A, Maldonado D, González JA, Molina D, Palacios E, Cantú P. Antiwear and extreme pressure properties of nanofluids for industrial applications. Tribology Transactions. 2014;57:1072–76.

[50] Rahmati B, Sarhan AA, Sayuti M. Morphology of surface generated by end milling AL6061-T6 using molybdenum disulfide (MoS2) nanolubrication in end milling machining. Journal of Cleaner Production. 2014;66:685–91.

[51] Sia SY, Bassyony EZ Sarhan AAD. Development of SiO2 nanolubrication system to be used in sliding bearings. International Journal of Advanced Manufacturing Technology. 2014;71(5–8):1277–84.

[52] Rapoport L, Lvovsky M, Lapsker I, Leshchinsky V, Volovik Y, Feldman Y, Margolin A, Rosentsveig R, Tenne R Slow release of fullerene-like WS2 nanoparticles from Fe–Ni graphite matrix: A self-lubricating nanocomposite. Nano Letters. 2001;1(3):137–40.

[53] Kim HS, Park JW, Park SM, Lee JS, Lee YZ Tribological characteristics of paraffin liquid with nanodiamond based on the scuffing life and wear amount. Wear. 2013;301(1–2):763–67.

[54] Liu LN, Fang ZP, Gu AJ, Guo ZH Lubrication effect of the paraffin oil filled with functionalized multiwalled carbon nanotubes for bismaleimide resin. Tribology Letters. 2011;42(1):59–65.

[55] Sui TY, Song BY, Zhang F, Yang QX. Effect of particle size and ligand on the tribological properties of amino functionalized hairy silica nanoparticles as an additive to polyalphaolefin. Journal of Nanomaterials. 2015;2015:492401.

[56] Bao YY, Sun JL, Kong LH. Tribological properties and lubricating mechanism of SiO2 nanoparticles in water-based fluid. IOP Conference Series: Materials Science and Engineering. 2017;182:012025.

[57] Kheireddin BA, Lu W, Chen I-C, Akbulut M. Inorganic nanoparticle-based ionic liquid lubricants. Wear. 2013;303:185–90.

[58] Konicek AR, Jacobs PW, Webster MN, Schilowitz AM. Role of tribofilms in wear protection. Tribology International. 2016;94:14–19.

[59] Gustavsson F, Jacobson S. Diverse mechanisms of friction induced self-organisation into a low-friction material – An overview of WS2 tribofilm formation. Tribology International. 2016;101:340–47.

[60] Alimirzaloo V, Qaleh SSG, Keshtiban PM, Ahmadi S. Investigation of the effect of CuO and Al2O3 nanolubricants on the surface roughness in the forging process of aluminum alloy. Proceedings of the Institution of Mechanical Engineers, Part J. 2017;231:1595–604.

[61] Rahmati B, Sarhan AA, Sayuti M. Morphology of surface generated by end milling AL6061-T6 using molybdenum disulfide (MoS2) nanolubrication in end milling machining. Journal of Cleaner Production. 2014;66:685–91.

[62] Kheireddin BA, Lu W, Chen I-C, Akbulut M. Inorganic nanoparticle-based ionic liquid lubricants. Wear. 2013;303:185–90.

[63] Kim HS, Park JW, Park SM, Lee JS, Lee YZ Tribological characteristics of paraffin liquid with nanodiamond based on the scuffing life and wear amount. Wear. 2013;301(1–2):763–67.

[64] Singh A, Chauhan P, Mamatha TG. A review on tribological performance of lubricants with nanoparticles additives. Materials Today: Proceedings. 2020;25:586–91.

[65] Luo T, Wei X, Zhao H, Cai G, Zheng X. Tribology properties of Al2O3/TiO2 nanocomposites as lubricant additives. Ceramics International. 2014;40:10103–09.

[66] Ali MKA, Fuming P, Younus HA, Abdelkareem MAA, Essa FA, Elagouz A, Xianjun H. Fuel economy in gasoline engines using Al2O3/TiO2 nanomaterials as nanolubricant additives. Applied Energy. 2018;211:461–78.

[67] Kumar N, Bhaumik S, Sen A, Shukla AP, Pathak SD. One-pot synthesis and first-principles elasticity analysis of polymorphic MnO2 nanorods for tribological assessment as friction modifiers. RSC Advances. 2017;7:34138–48.

[68] Uflyand IE, Zhinzhilo VA, Burlakova VE. Metal-containing nanomaterials as lubricant additives: State-of-the-art and future development. Friction. 2019;7(2):93–116. doi: 10.1007/s40544-019-0261-y.

[69] Peña-Parás L, Taha-Tijerina J, Garza L, Maldonado-Cortés D, Michalczewski R, Lapray C. Effect of CuO and Al2O3 nanoparticle additives on the tribological behavior of fully formulated oils. Wear. 2015;332–333:1256–61.

[70] He Z, Que W. Molybdenum disulfide nanomaterials: Structures, properties, synthesis and recent progress on hydrogen evolution reaction. Applied Materials Today. 2016;3:23–56. doi: 10.1016/j.apmt.2016.02.001.
[71] Padgurskas J, Rukuiza R, Prosyčevas I, Kreivaitis R. Tribological properties of lubricant additives of Fe, Cu and Co nanoparticles. Tribology International. 2013;60:224–32. doi: 10.1016/j.triboint.2012.10.024.
[72] Asrul M, Zulkifli NWM, Masjuki HH, Kalam MA. Tribological properties and lubricant mechanism of Nanoparticle in Engine Oil. Procedia Engineering. 2013;68:320–25. doi: 10.1016/j.proeng.2013.12.186.
[73] Azman NF, Samion S, Mnhm S, Sot MNHM. Investigation of tribological properties of CuO/palm oil nanolubricant using pin-on-disc tribometer. Green Mater. 2018;6(1):30–37. doi: 10.1680/jgrma.17.00026.
[74] Koshy CP, Rajendrakumar PK, Thottackkad MV. Evaluation of the tribological and thermo-physical properties of coconut oil added with MoS2 nanoparticles at elevated temperatures. Wear. 2015;330–331:288–308.
[75] Alves SM, Barros BS, Trajano MF, Ribeiro KSB, Moura E. Tribological behavior of vegetable oil-based lubricants with nanoparticles of oxides in boundary lubrication conditions. Tribology International. 2013;65:28–36. doi: 10.1016/j.triboint.2013.03.027.
[76] Qi M, Xiao J, Gong C, Jiang A, Chen Y. Evolution of the mechanical and tribological properties of DLC thin films doped with low-concentration hafnium on 316L steel. Journal of Physics D. 2018;51(2):025301. doi: 10.1088/1361-6463/aa9c2d.
[77] Ghaednia H, Jackson RL, Khodadadi JM. Experimental analysis of stable CuO nanoparticle enhanced lubricants. Journal of Experimental Nanoscience. 2015;10(1):1–18. doi: 10.1080/17458080.2013.778424.
[78] Ghaednia H, Jackson RL. The effect of nanoparticles on the real area of contact, friction and wear. Journal of Tribology. 2013;135(4):041603. doi: 10.1115/1.4024297.
[79] Spear JC, Ewers BW, Batteas JD. 2D-nanomaterials for controlling friction and wear at interfaces. Nano Today. 2015;10(3):301–14. doi: 10.1016/j.nantod.2015.04.003.
[80] Gonzalez-Rodriguez P, Van Den Nieuwenhuijzen KJH, Lette W, Schipper DJ, Ten Elshof JE. Tribochemistry of bismuth and bismuth salts for solid lubrication. ACS Applied Materials & Interfaces. 2016;8(11):7601–06. doi: 10.1021/acsami.6b02541,. PMID 26936490
[81] Flores-Castañeda M, Camps E, Camacho-López M, Muhl S, García E, Figueroa M. Bismuth nanoparticles synthesized by laser ablation in lubricant oils for tribological tests. Journal of Alloys and Compounds. 2015;643:S67–70. doi: 10.1016/j.jallcom.2014.12.054.
[82] Scherge M, Böttcher R, Kürten D, Linsler D. Multi-Phase friction and wear reduction by copper nanopartices. Lubricants. 2016;4(4):36. doi: 10.3390/lubricants4040036.
[83] Wang XL, Yin YL, Zhang GN, Wang WY, Zhao KK. Study on antiwear and repairing performances about mass of nano-copper lubricating additives to 45 Steel. Physics Procedia. 2013;50:466–72. doi: 10.1016/j.phpro.2013.11.073.
[84] Padgurskas J, Rukuiza R, Prosyčevas I, Kreivaitis R. Tribological properties of lubricant additives of Fe, Cu and Co nanoparticles. Tribology International. 2013;60:224–32. doi: 10.1016/j.triboint.2012.10.024.
[85] Zhang SW, Hu LT, Feng DP, Wang HZ. Anti-wear and friction-reduction mechanism of Sn and Fe nanoparticles as additives of multialkylated cyclopentanes under vacuum condition. Vacuum. 2013;87:75–80. doi: 10.1016/j.vacuum.2012.07.009.
[86] Sánchez-López JC, Abad MD, Kolodziejczyk L, Guerrero E, Fernández A. Surface-modified Pd and Au nanoparticles for anti-wear applications. Tribology International. 2011;44(6):720–26. doi: 10.1016/j.triboint.2009.12.013.
[87] Abad MD, Sánchez-López JC. Tribological properties of surface-modified Pd nanoparticles for electrical contacts. Wear. 2013;297(1–2):943–51. doi: 10.1016/j.wear.2012.11.009.
[88] Chou R, Battez AH, Cabello JJ, Viesca JL, Osorio A, Sagastume A. Tribological behavior of polyalphaolefin with the addition of nickel nanoparticles. Tribology International. 2010;43(12):2327–32. doi: 10.1016/j.triboint.2010.08.006.

[89] Xue WG, Zhao ZH, Wang P, Jin ZL, Xu XH, Zhou XG. Performance study of zinc oxide nanoparticles for lubricant oil. Advanced Materials Research. 2015;1118:195–204. doi: 10.4028/www.scientific.net/AMR.1118.195.

[90] Harta IC, Owens K, De Jesús Santiago S, Schall D, Thrush S. Barber Get al.Tribologicalperformance of ZnO–oil nanofluids at elevatedtemperatures. SAE International Journal of Fuels and Lubricants. 2013;6(1):126–31. doi: 10.4271/2013-01-1219.

[91] Chang H, Li ZY, Kao MJ, Huang KD, Wu HM. Tribological property of TiO2nanolubricant on piston and cylinder surfaces. Journal of Alloys and Compounds. 2010;495(2):481–84. doi: 10.1016/j.jallcom.2009.10.017.

[92] Arumugam S, Sriram G. Preliminary Study of Nano- and microscale TiO2 additives on tribological behavior of chemically modified rapeseed oil. Tribology Transactions. 2013;56(5):797–805. doi: 10.1080/10402004.2013.792977.

[93] Sharma AK, Tiwari AK, Singh RK, Dixit AR. Tribological Investigation of TiO2 nanoparticle based cutting fluid in machining under minimum quantity lubrication (MQL). Materials Today: Proceedings. 2016;3(6):2155–62. doi: 10.1016/j.matpr.2016.04.121.

[94] Ma SY, Zheng SH, Cao DX, Guo HN. Anti-wear and friction performance of ZrO2 nanoparticles as lubricant additive. Particuology. 2010;8(5):468–72. doi: 10.1016/j.partic.2009.06.007.

[95] Wu J, Huang XZ, Berglund K, Lu XH, Feng X. Larsson Ret al.CuO nanosheets produced in graphene oxide solution: An excellent anti-wear additive for self-lubricating polymer composites. Composites Science and Technology. 2018;162:86–92. doi: 10.1016/j.compscitech.2018.04.020.

[96] An V, Irtegov Y, Izarra Cd. Study of Tribological properties of nanolamellarWS2 and MoS2 as additives to lubricants. Journal of Nanomaterials. 2014;2014:865839.

[97] Yi MR, Zhang CH. The synthesis of MoS2 particles with different morphologies for tribological applications. Tribology International. 2017;116:285–94. doi: 10.1016/j.triboint.2017.06.045.

[98] Hu KH, Hu XG, Xu YF, Huang F, Liu JS. The Effect of morphology on the tribological properties of MoS2 in liquid paraffin. Tribology Letters. 2010;40(1):155–65. doi: 10.1007/s11249-010-9651-z.

[99] Wang HD, Liu YH, Chen Z, Wu BB, Xu SL, Luo JB. Layered double hydroxide nanoplatelets with excellent tribological properties under high contact pressure as water-based lubricant additives. Scientific Reports. 2016;6:22748. doi: 10.1038/srep22748. PMID 26951794

[100] Wang HD, Liu YH, Guo FM, Sheng HP, Xia KL. LiuWRetal.Catalytically active oil-based lubricant additives enabled by calcining Ni–Al layered double hydroxides. Journal of Physical Chemistry Letters. 2020;11(1):113–20. doi: 10.1021/acs.jpclett.9b03094,. PMID 31823614

[101] Zhao FY, Bai ZM, Fu Y, Zhao D, Yan CM. Tribological properties of serpentine, La (OH)3 and their composite particles as lubricant additives. Wear. 2012;288:72–77. doi: 10.1016/j.wear.2012.02.009.

[102] Zhang M, Wang XB, Fu XS, Xia YQ. Performance and anti-wear mechanism of CaCO3 nanoparticles as a green additive in poly-alpha-olefin. Tribology International. 2009;42(7):1029–39. doi: 10.1016/j.triboint.2009.02.012.

[103] He XL, Xiao HP, Choi H, Díaz A, Mosby B. Clearfield Aet al.α-zirconium phosphate nanoplatelets as lubricant additives. Colloids and Surfaces A: Physicochemical and Engineering Aspects. 2014;452:32–38. doi: 10.1016/j.colsurfa.2014.03.041.

[104] Li JS, Hao LF, Xu XH, Zhang L, Ren TH. Preparation, characterization and tribological evaluation of calcium borate nanoparticles as lubricant additives. Advanced Materials Research. 2010;148–149:1047–56. doi: 10.4028/www.scientific.net/AMR.148-149.1047.

[105] Lv YX, Zhao XC, Cheng JW, Liu Y, Li H. The study on the reducing friction and anti-wear performance of zinc phosphate nanoparticles as lubrication additives. Materials Science Forum. 2013;749:211–16. doi: 10.4028/www.scientific.net/MSF.749.211.

[106] Li ZW, Hou X, Yu LG, Zhang ZJ, Zhang PY. Preparation of lanthanum trifluoride nanoparticles surface-capped by tributyl phosphate and evaluation of their tribological properties as lubricant additive in liquid paraffin. Applied Surface Science. 2014;292:971–77. doi: 10.1016/j.apsusc.2013.12.089.

[107] Yang HM, Li JS, Zeng XQ. Tribological behavior of nanocarbon materials with different dimensions in aqueous systems. Friction. 2020;8(1):29–46. doi: 10.1007/s40544-018-0235-5.

[108] Joly-Pottuz L, Vacher B, Ohmae N, Martin JM, Epicier T. Anti-wear and friction reducing mechanisms of carbon nano-onions as lubricant additives. Tribology Letters. 2008;30(1):69–80. doi: 10.1007/s11249-008-9316-3.

[109] Liu LN, Fang ZP, Gu AJ, Guo ZH. Lubrication effect of the paraffin oil filled with functionalized multiwalled carbon nanotubes for bismaleimide resin. Tribology Letters. 2011;42(1):59–65. doi: 10.1007/s11249-011-9749-y.

[110] Kumar N, Saini V, Bijwe J. Performance properties of lithium greases with PTFE particles as additive: Controlling parameter-size or shape?. Tribology International. 2020;148. doi: 10.1016/j.triboint.2020.106302.

[111] Wang JM, Wang JH, Li CS, Zhao GQ, Wang XB. Tribological performance of poly (sodium 4-styrenesulphonate) as additive in water-glycol hydraulic fluid. Lubrication Science. 2012;24(3):140–51. doi: 10.1002/ls.1169.

[112] Tu TT, Wang WX, Wan Y, Pu JB. Tribology properties of the aqueous solution of polyvinylpyrrolidone. Materials Science. 2014;4(3):103–10.

[113] Eswaraiah V, Sankaranarayanan V, Ramaprabhu S. Graphene-based engine oil nanofluids for tribological applications. ACS Applied Materials & Interfaces. 2011;3(11):4221–27. doi: 10.1021/am200851z. PMID 21981026

[114] Mungse HP, Khatri OP. Chemically functionalized reduced graphene oxide as a novel material for reduction of friction and wear. Journal of Physical Chemistry C. 2014;118(26):14394–402. doi: 10.1021/jp5033614.

[115] Schlüter B, Mülhaupt R, Kailer A. Synthesis and tribological characterization of stable dispersions of thermally reduced graphite oxide. Tribology Letters. 2014;53(1):353–63. doi: 10.1007/s11249-013-0275-y.

[116] Zhao J, Li YR, Wang YF, Mao JY, He YY, Luo JB. Mild thermal reduction of graphene oxide as a lubrication additive for friction and wear reduction. RSC Advances. 2017;7(3):1766–70. doi: 10.1039/C6RA26488C.

[117] Zhao J, Li YR, Mao JY, He YY, Luo JB. Synthesis of thermally reduced graphite oxide in sulfuric acid and its application as an efficient lubrication additive. Tribology International. 2017;116:303–09. doi: 10.1016/j.triboint.2017.06.023.

[118] Zhang CL, Zhang SM, Yu LG, Zhang ZJ, Wu ZS, Zhang PY. Preparation and tribological properties of water-soluble copper/silica nanocomposite as a water-based lubricant additive. Nanocomposite-based lubricant additives. Applied Surface Science. 2012;259:824–30.

[119] Luo T, Wei XW, Zhao HY, Cai GY, Zheng XY. Tribology properties of Al2O3/TiO2 nanocomposites as lubricant additives. Ceramics International. 2014;40(7):10103–09. doi: 10.1016/j.ceramint.2014.03.181.

[120] Li XF, Peng SX. Tribology of poly (vinyl alcohol)-carbon nanotube composites. In: Proceedings of the 2011 International Conference on Electronic & Mechanical Engineering and Information Technology. Harbin, China, 2011, 1431–34. doi: 10.1109/EMEIT.2011.6023316.

[121] Yu B, Liu ZL, Zhou F, Liu WM, Liang YM. A novel lubricant additive based on carbon nanotubes for ionic liquids. Materials Letters. 2008;62(17–18):2967–69. doi: 10.1016/j.matlet.2008.01.128.

[122] Jia XH, Huang J, Li Y, Yang J, Song HJ. Monodisperse Cunanoparticles@MoS2 nanosheets as a lubricant additive for improved tribological properties. Applied Surface Science. 2019;494:430–39. doi: 10.1016/j.apsusc.2019.07.194.

[123] Song W, Yan JC, Ji HB. Fabrication of GNS/MoS2 composite with different morphology and its tribological performance as a lubricant additive. Applied Surface Science. 2019;469:226–35. doi: 10.1016/j.apsusc.2018.10.266.

[124] Zhang MS, Chen BB, Tang H, Tang GG, Li CS. Chen Let al.Hydrothermal synthesis and tribological properties of FeS2 (pyrite)/reduced graphene oxide heterojunction. RSC Advances. 2015;5(2):1417–23. doi: 10.1039/C4RA09261A.

[125] Meng Y, Su FH, Chen YZ. Supercritical fluid synthesis and tribological applications of silver nanoparticle-decorated graphene in engine oil nanofluid. Scientific Reports. 2016;6:31246. doi: 10.1038/srep31246. PMID 27488733

[126] Wang L, Gong PW, Li W, Luo T, Cao BQ. Mono-dispersed Ag/graphene nanocomposite as lubricant additive to reduce friction and wear. Tribology International. 2020;146. doi: 10.1016/j.triboint.2020.106228.

[127] Meng Y, Su FH, Chen YZ. Synthesis of Nano-Cu/graphene oxide composites by supercritical CO2-assisted deposition as a novel material for reducing friction and wear. Chemical Engineering Journal. 2015;281:11–19. doi: 10.1016/j.cej.2015.06.073.

[128] Zhao J, Li YR, He YY, Luo JB. In situ green synthesis of the new sandwich like nanostructure of Mn3O4/graphene as lubricant additives. ACS Applied Materials & Interfaces. 2019;11(40):36931–38. doi: 10.1021/acsami.9b08993,. PMID 31356745

[129] Kogovšek J, Kalin M. Various MoS2-, WS2- and C-based micro- and nanoparticles in boundary lubrication. Tribology Letters. 2014;53(3):585–97.

[130] Zhao J, Huang YY, He YY, Shi YJ. Nanolubricant additives: A review. Friction. 2021;9(5):891–917. doi: 10.1007/s40544-020-0450-8.

[131] Duan LL, Li H, Zhang YT. Synthesis of hybrid nanoflower based carbonic anhydrase for enhanced biocatalytic activity and stability. ACS Omega. 2018;3(12):18234–41. doi: 10.1021/acsomega.8b02247.

[132] Duan LL, Wang HX, Liu JD, Zhang YT. Three-dimensional self-assembled graphene oxide/enzymee in the presence of copper phosphate. Biomedical Physics and Engineering Express. 2015;1(4):045101. doi: 10.1088/2057-1976/1/4/045101.

[133] Spear JC, Ewers BW, Batteas JD. 2D-nanomaterials for controlling friction and wear at interfaces. Nano Today. 2015;10(3):301–14. doi: 10.1016/j.nantod.2015.04.003.

[134] Jin YL, Li J, Cheng BX, Jia D, Tu JS, Zhan S, et al. Thermal oxidation behavior of trimethylolpropane trioleate base oil when exposed to iron surfaces. Industrial Lubrication and Tribology. 2019;72 (3):473–78. doi: 10.1108/ILT-08-2019-0329.

[135] Wang SZ, McGuirk CM, Ross MB, Wang SY, Chen PC, Xing Het, et al. General and direct method for preparing oligonucleotide-functionalized metal–organic framework nanoparticles. Journal of the American Chemical Society. 2017;139(29):9827–30. doi: 10.1021/jacs.7b05633. PMID 28718644.

[136] Bhattacharjee S, Jang MS, Kwon HJ, Ahn WS. Zeolitic imidazolate frameworks: Synthesis, functionalization, and catalytic/adsorption applications. Catalysis Surveys from Asia. 2014;18(4):101–27. doi: 10.1007/s10563-014-9169-8.

[137] Yu L, Zhang L, Ye F, Sun M, Cheng X, Diao G. Preparation and tribological properties of surface-modified nano-Y2O3 as additive in liquid paraffin. Applied Surface Science. 2012;263:655–59.

[138] Gong K, Lou W, Zhao G, Wu X, Wang X. MoS 2 nanoparticles grown on carbon nanomaterials for lubricating oil additives. Friction. 2021 Aug;9:747–57.

[139] Bojarska Z, Kopytowski J, Mazurkiewicz-Pawlicka M, Bazarnik P, Gierlotka S, Rożeń A, Makowski Ł. Molybdenum disulfide-based hybrid materials as new types of oil additives with enhanced tribological and rheological properties. Tribology International. 2021 Aug 1;160:106999.

[140] Tomala A, Ripoll MR, Kogovšek J, Kalin M, Bednarska A, Michalczewski R, Szczerek M. Synergisms and antagonisms between MoS2 nanotubes and representative oil additives under various contact conditions. Tribology International. 2019 Jan 1;129:137–50.

[141] Duan L, Li J, Duan H. Nanomaterials for lubricating oil application: A review. Friction. 2023 May;11(5):647–84.

Index

https://doi.org/10.1515/9783111145136-013

Also of interest

Optimized Equipment Lubrication.
Conventional Lube, Oil Mist Technology and Full Standby Protection
Heinz Bloch, 2021
ISBN978-3-11-074934-2, e-ISBN (PDF)978-3-11-074944-1

Essential Oils.
Sources, Production and Applications
Edited by: Rajendra Chandra Padalia, Dakeshwar Kumar Verma, Charu Arora and Pramod Kumar Mahish, 2023
ISBN 978-3-11-079159-4, e-ISBN (PDF) 978-3-11-079160-0

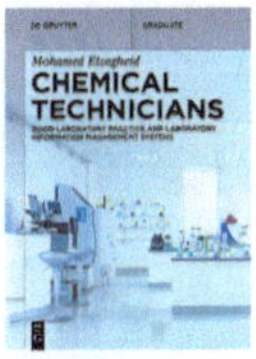

Chemical Technicians.
Good Laboratory Practice and Laboratory Information Management Systems
Mohamed Elzagheid, 2023
ISBN 978-3-11-119110-2, e-ISBN (PDF) 978-3-11-119149-2

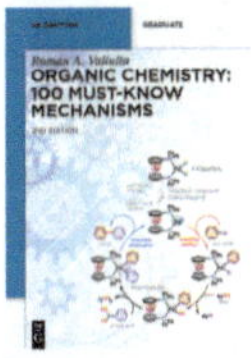

Organic Chemistry: 100 Must-Know Mechanisms
Roman Valiulin, 2023
ISBN 978-3-11-078682-8, e-ISBN (PDF) 978-3-11-078683-5

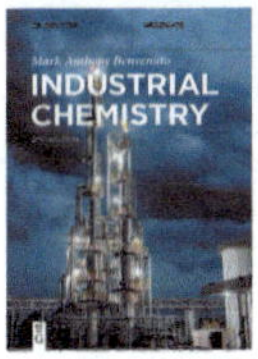

Industrial Chemistry
Mark Anthony Benvenuto, 2023
ISBN 978-3-11-067106-3, e-ISBN (PDF) 978-3-11-067109-4